Lecture Notes in Mathematics

History of Mathematics Subseries

Volume 2278

Series Editor

Patrick Popescu-Pampu, CNRS, UMR 8524 - Laboratoire Paul Painlevé, Université de Lille, Lille, France

Providing captivating insights into facets of the recent history of mathematics, the volumes in this subseries of LNM explore interesting developments of the past 200 years or so of research in this science. Their aim is to emphasize the evolution of its intellectual discourse, the emergence of new concepts and problems, the ground-breaking innovations, the human interactions, and the surrounding events that all contributed to weave the backdrop of today's research and teaching in mathematics. These high-level and largely informal accounts will be of interest to researchers and graduate students in the mathematical sciences, in the history or the philosophy of mathematics, and to anyone who seeks to understand the historical growth of the discipline.

More information about this subseries at http://www.springer.com/series/8909

Frédéric Patras

The Essence of Numbers

 Springer

Frédéric Patras (ID)
CNRS
Université Côte d'Azur
Nice, France

ISSN 0075-8434   ISSN 1617-9692   (electronic)
Lecture Notes in Mathematics
ISSN 2193-1771   ISSN 2625-7157   (electronic)
History of Mathematics Subseries
ISBN 978-3-030-56699-9   ISBN 978-3-030-56700-2   (eBook)
https://doi.org/10.1007/978-3-030-56700-2

Mathematics Subject Classification: 00A30, 03E10, 01A99, 03A05

Translation from the French language edition: *La possibilité des nombres* by Fréderic Patras, © Presses Universitaires de France 2014. Published by Presses Universitaires de France. All Rights Reserved.

This Springer imprint is published by the registered company Springer Nature Switzerland AG.
The registered company address is: Gewerbestrasse 11, 6330 Cham, Switzerland

Preface to the English Translation

This book originates in a lecture series on the philosophy of arithmetic given for students in philosophy at the University of Nice. One of the aims of the lectures was to let a mathematician introduce them to the philosophy of science and of mathematics. I have indeed been practising philosophy, its reading, writing, and have been attending and giving academic seminars for a very long time, since the early 1990s, but have remained a mathematician by training, profession and theoretical interests.

These interests may sometimes differ from those of a philosopher of science. Fortunately, the boundaries between mathematics and philosophy are evolving, with much more attention recently paid by philosophers to the actual practice of mathematics. A much wider spectrum of approaches is now developing at the boundary of the two disciplines: the traditional interest in logic, language and foundations has recently been augmented by a large scale rediscovery and interest in the history of thought, ontology, metaphysics, mathematical practice, Platonism... My personal conviction, illustrated in this book, is that allowing such a broad spectrum of approaches to mathematics fits with mathematical experience better than such and such unilateral view on what mathematics should be.

Turning back to the lecture series, their initial aim was to focus on what I consider one of the deepest texts ever written on mathematical thought, Husserl's *Philosophy of Arithmetic*. However, I realized almost immediately that the subject was touching on so many topics and authors—central to both mathematics and philosophy—that the scope had to be extended. Most references I could find focused on specific features of numbers and their history. Moreover, they did not reflect or explain what I found most intellectually interesting, exciting or challenging in authors such as Aristotle, Husserl, Plato or Frege on numbers.

Progressively, as the book evolved, it began to cover debates on the nature of numbers from the Presocratics to the modern period. Its targeted audience was augmented to include, besides philosophers, mathematicians interested in philosophy. Hopefully it will be useful for a variety of readers, from scholars who have a general interest in philosophy or mathematics to philosophers and mathematicians themselves.

One of the features that make the book different is that history, mathematics and philosophy, instead of being pursued for themselves, are used as tools to account for the essence of numbers. The subject pervades so many different fields, from our daily experience to debates still active in philosophy or mathematical logic, that adopting a single-minded point of view would anyway have obscured most of its interesting features and hidden its intrinsic complexity. Accordingly, this volume has several purposes. First of all, it serves as an introduction to mathematical thought in all its complexity, showing that even the natural numbers, the simplest and most common mathematical objects, have a very complex nature whose study leads to understanding subtle features of the functioning of our thought. Notice that, by the complex nature of numbers, we do not mean complex arithmetical properties—say, Riemann's conjecture—but the complexity hidden in the understanding of zero, one, the notions of collection or of set, of induction, of infinity. We are all familiar with these objects and notions, and this familiarity obscures the underlying principles that can also be found in the much more abstract objects and theories of contemporary mathematics.

It also serves as an introduction to the philosophy of mathematics, and it does so by showing how a given problem (say, the understanding of cardinal numbers) can lead to the development of various theoretical approaches that might intersect or complement each other. In this example of cardinal numbers, metaphysics, logic, set theory, the study of language, jointly contributed to their understanding at the end of the nineteenth century, paving the way for some of the most important later developments in mathematical logic and twentieth century philosophy.

Lastly, this book will serve as an introduction to the works of major mathematicians and philosophers, from Plato and Aristotle to Cantor, Dedekind, Frege, Husserl and Weyl. The adopted point of view deviates from the classical and dominant one in the literature. The account of Frege, for example, is largely different from the common one in philosophy that features his contributions to the philosophy of language and more generally to analytic philosophy, often losing sight of the properly mathematical content of his thought.

If one wanted to assign this book to a specific category, it would be historical epistemology, which allows one to consider scientific objects and concepts in the context of their production and genesis, featuring their epistemological relevance. Numbers as they are treated in this book are indeed epistemic objects: mathematical objects that have been subject to epistemological inquiry and attention throughout their history and whose understanding has evolved accordingly. Historical epistemology jointly uses history, philosophy and science to grasp the content of concepts, their emergence, variations and transformations. Surprisingly enough, numbers, seemingly the simplest and best-known mathematical entities, can be the subject of such analyses.

That being said, this book does not make any claim to address such methodological questions. Its sole and unique scope is the understanding of natural numbers and the manifold possible approaches to it, past and present.

The present volume is a translation of the French edition, *La possibilité des nombres*, Presses Universitaires de France, 2014. I thank the publisher thereof, Humensis, for allowing me to publish an English version. The content is essentially unchanged, except for minor modifications aiming at fluidifying the style and improving the presentation.

Nice, France   Frédéric Patras
May 2020

Contents

Chapter 1
Introduction

Number is, along with geometry, at the origin of mathematical thinking. From the most elementary counting activity to the uses made of it in contemporary theories, it is a universal concept par excellence, present both in everyday life and in the most advanced mathematical or logical debates. However, in spite of the evidence that accompanies its everyday use, its understanding is far from self-evident and presents two apparently contradictory characteristics. First of all, from the origins of civilization to the present day, so-called natural numbers (one, two, three...) have hardly evolved and our intuitive understanding of them has probably changed little since the Greece of Thales or Plato. However, their theorization within mathematics has progressed surprisingly, to the point of making the concept of number the arbiter of some of the most profound debates that animated mathematical thought at the beginning of the last century, with, to mention only the most prominent, Cantor's infinities, the paradoxes of set theory or, more recently, the aporia of mathematical logic and the works of Gödel.

This deep tension between the evidence, the immediacy of numbers and the possibility of looking at them from complex theoretical points of view makes their mathematical, historical and philosophical study rich in lessons. To understand the "essence of numbers" is therefore also to understand the springs, the mechanisms that govern scientific thought. More concretely, trying to grasp their nature implies going through the different interpretations that have been proposed thereof and through the ways in which the latter relate to projects that may be as much a matter of philosophy (as in Plato, Aristotle, Kant, Wittgenstein...) as of more resolutely mathematical theories (with Euclid's Elements, Cartesian geometry, Cantor's infinities, set theory...).

It is this plurality, this richness of possible approaches to the concept of number that this book seeks to account for, without preconceived notions about the paths to be favoured: mathematics, epistemology, history and philosophy will be used in turn to tackle the various problems posed by the existence of numbers. As the general idea is to systematically privilege the contents of thought over technicality

© The Editor(s) (if applicable) and The Author(s), under exclusive license to Springer Nature Switzerland AG 2020
F. Patras, *The Essence of Numbers*, Lecture Notes in Mathematics 2278,
https://doi.org/10.1007/978-3-030-56700-2_1

and exhaustiveness, it is conceived more as an introduction to mathematical thought, its successes and aporias, than as a systematic treatise on numbers and their history. Its purpose will appear as a watermark. It will be to show that mathematics, if it makes continuous and spectacular progress in the extension of its field, progresses simultaneously in the understanding it has of its springs and its foundations.

Number, since that is what it will be about, is therefore anything but a simple entity, contrary to what the ordinary practice of calculation or enumeration might suggest. Mathematicians and philosophers have used various detours to arrive at a rigorous definition, the most famous and best accepted of which, even today, dates back to the end of the nineteenth century and involves set theory. This theory only came into being at the price of many difficulties and, to justify the existence of the most ordinary numbers, one, two, three..., had to resort to abstract processes whose complexity seems to be out of proportion with that of the entities that were being defined.

This paradoxical situation does not go without recalling the classical definition of man as a "rational animal" and the humanist objections of a Montaigne or a Descartes: it is a question of defining a common notion (man), and one would like to use for this purpose two other notions, rather more complex and abstract (the animal, rationality)! The case of numbers raises comparable difficulties since one can legitimately wonder which concept could precede that of number. Poincaré insisted on this at the beginning of the twentieth century: there are strong reasons to suspect that those who seek to justify the existence and properties of numbers implicitly use the notion in their reasoning.

How far is Poincaré's objection admissible? Is this, then, the way science should proceed, and is it ultimately possible to define everything? There are several juxtaposed problems which make any simple answer unsatisfactory. Philosophy would distinguish here, in the old language of Aristotelian philosophy, between substantial anteriority and logical anteriority. In the order of reality and phenomena, and especially in everyday life (substantial anteriority), the knowledge of numbers largely pre-exists all attempts to define them, just as we know how to recognize a man long before we understand what a rational animal can be. In the architectural order of reason and the theory of knowledge (logical anteriority), this pre-existence is less obvious, and it is understandable that concepts such as the one of set can occupy the first rank. The question then is whether the choice of the concept of set is justified as a basis for arithmetic and whether other concepts could not just as legitimately play this role.

From this point of view, the contemporary period has something to teach. Indeed, recent theoretical and methodological advances have led to a profound re-evaluation of the way in which mathematical ideas emerge and crystallize. We now perceive the paths of scientific creation in a renewed way, and these upheavals are not without impact on our understanding of the fundamental mechanisms underlying the construction and use of numbers.

1.1   Greek Origins

Theoretical thought was born in Greece and Greek thought continues to structure our conception of science. Number played a decisive role in this birth by crystallizing in the problem of its origin, in Plato's time, some of the questions that the previous philosophies, the so-called presocratic ones, had begun to raise. The problem of the One, straddling between mathematics and metaphysics, has thus been able to traverse the history of thought, from Parmenides to the neo-Platonic theologies of the Middle Ages and beyond. The thesis that "one is not a number" is undoubtedly one of its most persistent and significant avatars. Although difficult to understand nowadays, the thesis was still discussed and debated in nineteenth-century mathematical literature. Among the great intuitions of Greek thought that are still at work in current thinking on mathematics and logic is the idea of a difference in nature between number and magnitude, which structures the fundamental opposition between arithmetic and geometry.

Beyond these few major themes, the major teaching of Greek thought is perhaps primarily the elevation of the idea of number to the rank of an epistemological problem. Of course, many civilizations preceded Greece in the use of numbers and calculation, but none seems to have been concerned with defining numbers and legitimizing these calculations. There is a genuine difficulty in understanding the very possibility of such a questioning, and it took all the Greek genius to conceive of it and measure its necessity. After all, the usual rules for counting, measuring and calculating are, in practice, quite sufficient, and the opportunity to reflect on the origin of these rules is not obvious. The same is true, moreover, of all the concepts on which our daily judging activity is based: law, justice, laws, government, to name but a few. Our lives could very well pass without us ever having to think seriously about the content they implicitly convey. A mathematician may very well work without ever having to reflect on the origin of the concepts he manipulates, and this is even the normal way science should proceed, as it cannot constantly go back to its foundations. A strong determination and a great intellectual exigency were therefore necessary for Greek thought to free itself from the straitjacket of daily evidence and to problematize knowledge and its methods. The birth of theoretical thought and, incidentally, of the philosophy of arithmetic,[1] was at this price.

1.2   The Contribution of Mathematics

As decisive as Greek thought was in the constitution of a theory of scientific knowledge, the understanding of the idea of number remains inseparable from the

---

[1] We will call "philosophy of arithmetic" the study of the general idea of number rather than the study of the field of so-called "arithmetic" phenomena in modern mathematics (diophantine equations, algebraic number theory. . .).

historical progress of mathematical knowledge. The concepts that take shape in mathematical practice often shed light in turn on their methodological foundations, and a general reflection on the nature of the objects and springs of scientific thought is inseparable from its technical advances. In this sense, and although mathematical truth is timeless, the philosophy of science is largely dependent on its history or, more precisely, on a certain form of historicity of thought, since the very modalities of emergence of new ideas or results have an intrinsic theoretical significance.

To take just one example, the study of the arithmetic properties of the continuum provides an illustration that runs through the whole of mathematical history and has regularly renewed the terms of the relationship between arithmetic and geometry. It first led to the discovery of various natural generalizations of positive integers and their ratios (the fractions): square and cubic roots, irrational numbers,[2] and to the discovery of Archimedes' axiom[3] which guarantees the homogeneity and compatibility of continuum measurements. Closer to us, the study of the continuum at the end of the nineteenth century with Dedekind and Cantor was one of the driving forces behind the change of perspective leading to the privileging, as a basis for mathematics, of numbers over space. These are textbook examples of situations in which mathematical progress calls into question an entire theoretical edifice.

The difference between number and magnitude has undoubtedly played a key role here, because it captures the relationship of number to extent and, beyond that, to our intuition of space and time. These phenomena, already difficult to think about in the technically limited context of Greek mathematics, have become increasingly problematic over the centuries as each discovery has shifted their contours. Thus, the algebraization of geometry in the seventeenth century contributed to dissolve the conceptual autonomy of space in the infinite potentialities of calculation. Later, the discovery of geometrical representations of complex numbers gave them legitimacy and a concrete existence, while further reinforcing an impression of permeability between what is space and what is number. These conceptual and technical shifts disturb and enrich mathematical philosophy. They sometimes lead to giving new legitimacy to ideas that had fallen into disuse and had long been considered outdated.

1.3   Gottlob Frege

The end of the nineteenth century will occupy a decisive place in this work, because no other period has contributed so much to the mathematical understanding of numbers. The resulting conception of number, which has become paradigmatic, is

---

[2]A number such as π or the square root of 2 is said to be irrational: it cannot be expressed as a ratio of two integers.

[3]Archimedes' axiom states that, given two (non-zero, positive) quantities, A and B, there is always an integer multiple of A greater than B.

based on set theory. However, it leaves a feeling of incompleteness that is difficult to understand without going back to the work of Gottlob Frege, central to all twentieth-century mathematical thought, but of which mathematical epistemology has much too often retained only the most consensual aspects.

It is in Frege's thought that the destiny of the modern idea of number, and much more, was played out. Frege's work presents two faces simultaneously. Firstly, it is part of a great philosophical tradition. Kant's work was one of the main origins of Frege's. Even if the latter was a break with Kantism, that he intended to renew, all of Frege's mathematical work was organized around classical philosophical notions: analytical and synthetic truths; a priori; concept and object. One of Frege's great programmatic ideas was to bring arithmetic back to the pure laws of thought, and thus to make arithmetic truths into truths analytically deduced from first principles structuring all possible forms of theoretical knowledge. In this, Frege's project was resolutely epistemological and philosophical and inseparable from a global reflection on the nature and the springs of scientific thought. The history and philosophy of science have largely ignored this dimension and have above all retained the other face of Frege's thinking, namely the concrete results of his research programme, such as the possibility of a logical formalization of the foundations of mathematics or the first developments of set theory.

As far as the philosophy of arithmetic is concerned, the Fregean contribution goes far beyond these technical developments. The former regained with Frege the scope and the breath that it had in antiquity, and the problem of defining numbers thus became again a decisive issue for the whole theory of knowledge. Frege understood from the outset that this definition, if it is to be radical, cannot dispense with a reflection on the processes of thought (logic), on the organization of scientific language (symbolism, syntax, grammar), or on the organic elements of discourse (concepts, objects).

If the Fregean work technically marked the entry into a new era for mathematical thought, its posterity has also been accompanied by a renewal of the very field of action of mathematical philosophy. Post-Fregean mathematical philosophy and logic, with in particular the works of Hilbert, Husserl and Gödel, thus brought with them a whole set of technical and conceptual elements, and new tools that made it possible to approach in a very original and mathematically deep way classical problems: does logic exhaust the idea of a system of laws of thought? Can mathematics be fully reconducted to logic and formalism? What is the mode of existence of mathematical objects?

1.4 From Arithmetic to Algebra

Another major problem, which a treatment of the idea of number cannot avoid, is finally superimposed on those considered so far. It concerns the intrinsically algebraic nature of numbers and arithmetic. At an elementary level, this translates into the possibility of extending the system of natural numbers. Such extensions do not go without methodological difficulties which, historically, have been overcome only with pain. At a more advanced level, this translates into the possibility of defining domains of numbers by purely algebraic and symbolic, or even categorical, procedures.

The late introduction of zero and negative quantities in arithmetic provides a striking illustration. The question of zero, for example, which is most often treated superficially and in a purely historical manner, raises very interesting difficulties of principle. If numbers measure quantities, it is only by extension that zero can be considered as a number in its own right. The habit of calculation is easily misleading, and the ease with which we can nowadays induce certain conceptual properties of zero from its operational legitimacy conceals epistemological difficulties which only belatedly found an acceptable solution.

It is significant to note here that a certain epistemological blindness (consisting in not even understanding that operating with zero can pose conceptual problems) results in easier access to calculations with zero which, in purely operational terms, indeed do not pose any difficulty, as all elementary school students know! This is a phenomenon that is encountered even at very advanced levels of mathematical thinking: it may very well happen that a mathematician is reluctant to engage in a calculation because he is not convinced of its methodological validity, for reasons that go beyond the technique itself. Such phenomena are indicative of complex modes of mathematical thinking that deserve to be examined, and for the study of which recent advances in the understanding of brain functioning provide interesting tools of analysis, for example by making it possible to distinguish different cerebral modes of arithmetic calculation.

Beyond the zero problem, and although natural numbers will remain the main thread of this book, we will go through other classical extensions of the number domain, trying to quickly identify their meaning and implications. Particular emphasis will be put on the intrinsically algebraic-formal dimension of number. The latter takes on a new meaning in the light of general mathematical procedures, implicit in some classical constructions of number but whose true scope has only recently been understood.

The significance of these ideas for the philosophy of arithmetic is complex. Each of the moments in the philosophy of arithmetic discussed, from the Greek period to the contemporary era, contributes to our understanding of numbers without any advance ever invalidating the previous points of view.

Nevertheless, some great ideas emerge, as we shall see, from this overview, and some great thoughts: those of Plato, Aristotle, Dedekind, Cantor, Frege, Hilbert, Gödel..., without whose frequentation it would be vain to claim to know today the nature of numbers.

Chapter 2
The Lasting Influence of Pythagorism

During the three millennia BC, Egyptian and Mesopotamian mathematics developed fairly advanced computational techniques. Although they did not address the problem of a conceptual number determination (what are numbers?), there is every reason to believe that numbers were then implicitly conceived as a property of numbered things. The fact that a number can be isolated from its material support is not evident, and the question will arise in the nineteenth century, when mathematicians will try to understand the exact nature of mathematical statements: why would the act of abstracting the number ten and the act of abstracting the colour white from the observation of a group of ten white marbles refer to two radically different types of experience and two radically different modes of conceptualization? Or, more abstractly and more generally, why and how would the nature of mathematical concepts be distinct from the nature of other concepts derived from experience?

In fact, it must be recognized that it is much easier to think of numbers as having a more autonomous existence with respect to the things they serve to enumerate than colour, whose existence is difficult to conceive of outside a material medium. The autonomy of the rules of calculation with regard to the things that are numbered further accentuates this idea of a specificity of mathematical concepts. Thus Egyptian and Mesopotamian mathematics already hinted at the possibility of a development of calculation that would be independent of the concrete meanings at stake: distribution of rations to an army, distribution of wheat...It is difficult to go beyond these few observations without advancing theses with uncertain conclusions; if algebraic calculus has its own logic and brings into play in its functioning formal structures that can easily be used retrospectively to interpret ancient texts, the recognition of these structures was indeed very long to be established.

The Greek theory of number contrasts with previous conceptions precisely because of its willingness to consider the nature of numbers beyond their roots in the practice of enumeration and calculation. The mathematical theory of number itself (arithmetic), the geometric theory of quantities, numerology (the mysticism of

© The Editor(s) (if applicable) and The Author(s), under exclusive license   7
to Springer Nature Switzerland AG 2020
F. Patras, *The Essence of Numbers*, Lecture Notes in Mathematics 2278,
https://doi.org/10.1007/978-3-030-56700-2_2

numbers), and the arithmetic features of philosophical questioning combine to form a complex and inseparable whole. These ideas were essential for the development of Greek thought as a whole, and of the later philosophical tradition. They continue, as we shall see, to influence our understanding of the role and meaning of mathematics.

The idea that number can be defined independently of its rules of empirical use represents a considerable advance. It implies a change in status: since number becomes an autonomous object of thought, it is possible to question the reason for its existence—still according to the characteristic intuition of Greek thought that one must distinguish between questions of fact and matters of principle, between the order of phenomena and the order of reasons. This process of reflection on numbers accompanies the birth of the idea of demonstration, which replaces non-theoretical practice and operational rules. It is to Thales (about 624–548 BC) that the first definition of number as well as the first proof are frequently attributed. A number would, according to him, be a "collection of units". In its elliptical character, this formula highlights a fundamental and immutable feature of the idea of number: a (cardinal) number is the result of the grouping together in a whole of entities of the same type, entities whose nature remains to be clarified, the units. To a certain extent, the whole philosophy of number, or philosophy of arithmetic[1] has been devoted since Thales to clarifying, deepening and discussing this first definition.

2.1   Numbers in the Pythagorean School and Numerology

The theorization proper begins with the Pythagorean school[2] and was accompanied by a true mysticism of numbers. The latter had a lasting influence; according to the testimony of Aristotle (384–322 BC):[3]

> Those known as the Pythagoreans were the first to devote themselves to mathematics and to advance it. Nourished in this discipline, they believed that the principles of mathematics are the principles of everything. And as of those principles numbers are by nature the first, and as in numbers the Pythagoreans believed that they saw a multitude of analogies with all that is and becomes; as they saw, moreover, that numbers express musical properties and proportions; as, finally, all other things seemed to them, in their entire nature, to be

---

[1]Recall that we use the term "philosophy of arithmetic" to refer to the study of the concepts of cardinal and natural numbers. The terminology, retained here for convenience, is obviously abusive: a philosophy of general arithmetic should also account for arithmetic phenomena such as the links between number theory, analysis and geometry.

[2]Pythagoras lived in the sixth century BC. The most famous arithmetic discovery of his school was the irrationality of the square root of 2.

[3]Translations of Aristotle, Plato and Plotinus available in the literature may differ and be subject to hermeneutic choices. In the present translation of *La Possibilité des nombres*, we have sometimes appealed to standard English translations but some other times, for consistency with the original version of the book, based the translation on the French one chosen there. Sometimes we also indicate alternative translations in a footnote. As the pagination refers to the standard edition of the texts, the reader can anyway refer easily to his or her preferred translation.

formed in the likeness of numbers, and that numbers seemed to be the primordial realities of the universe: in these conditions, they considered that the principles of numbers are the elements of all beings, and that the whole of Heaven is harmony and number.

Aristotle [6, A 5 985b]

From these determinations come surprising analogies. Thus, the soul, as a principle, was 1; intelligence, 2, a number representing the movement from the premises to the conclusion; justice, 4 or 9, square numbers representing perfect balance. As for number in all generality, some Pythagoreans define it as the "progression of a multiplicity beginning with a unity and a regression ending in it", a much more dynamic statement than Thales' definition. This definition corresponds to a recurring theme, an alternative to the one introduced by Thales, which anchors number in the ideas of succession, temporality and order.

The guiding features of Pythagorean numerology are not specific to it, and are shared by other currents of thought at the limit of philosophy and mysticism, as if it were one of the natural functions of numbers to underpin a certain form of esotericism. The ultimate reason probably lies in the very functioning of thought, whether rational or not, which needs logical and methodological reference points to build and develop itself, even when non-theoretical ways of thinking are involved. The thought of Lao Tzu (sixth century BC), which gave birth to Taoism and, later, to the Taoist religion, thus presents several features in common with presocratic philosophies and, in its relation to numbers, with Pythagorism.[4] In the terms of Western thought, with all that this implies of approximation to a radically different way of thinking, the Tao is the principle of all things, below Being and Non-Being. As such, it is inaccessible to discursive knowledge. However, there is a Taoist genesis of the world that can be described numerologically:

In chapter 42 of Lao Tzu, this genesis of the world is presented as follows: "Tao gave birth to One; One gave birth to Two; Two gave birth to Three; Three gave birth to ten thousand beings."

In this text where the cosmogony from the Tao to the formed beings is summarized, the numbers symbolize sub-principles and stages of genesis. We know how much the Chinese liked to use numbers to evoke, not quantities, but qualities. But here, it is at first sight surprising that the Tao gives birth to One, because doesn't One, a symbol of unity-totality, represent the Tao itself? [...]

This text can be glossed over with the *Huainanzi*.[5] It explains that the action of the Tao begins with Unity, but since Unity cannot give life, it is divided into Yin and Yang [...]. Two are Yin and Yang, but also Heaven and Earth; Three, the harmonious union of the previous ones, but also the measure of the rhythm of this union.

Kaltenmark [76]

Whatever the specificities of Taoism and its roots in the major categories of Chinese thought (Tao, Yin, Yang, five elements), the similarity of the role played by numerology in Taoism to that played in Pythagorism is striking and confirms the existence of a certain form of universality of the idea of number, which one could

---

[4]We follow here Max Kaltenmark [76].

[5]A collection of philosophical essays written in the second century BC.

seek to find in the various mysticisms, cosmologies or theogonies. One[6] is at the principle of things and beings, but, undivided, cannot have a generating function. It is with Two that the possibility of movement, of creation, of a dialectic appears. Three is the figure of the synthesis of the previous moments, constitutive of Two. Later, in Christianity, this conception of Three was to be placed at the service of a Trinitarian theology that is very surprising to whom ignores its origins.

The analogical use of number of Pythagorean inspiration has had a lasting influence on the theory of knowledge. It was with the Renaissance that it both threw its last fires and gradually lost all legitimacy, as the development of modern thought and its demand for rationality gradually banished the most debatable uses of analogy. The example of Charles de Bovelles (1478–1567) and his work, *Le Sage* [35], clearly illustrates these contradictions of Renaissance thought. It is an astonishing text, emblematic of this period of transition, in which resolutely humanistic and modern features coexist with surprising archaisms. In the judgment of Cassirer (1874–1945), one of the best theorists of Renaissance Humanism:

> [This is] perhaps the most remarkable creation and, in many respects, the most characteristic of Renaissance philosophy. Do we ever see the old and the new, the survivors and the creative forces, assembled in such a narrow space? [...] Suddenly, at the heart of a schematically allegorical exposition of the universe, thoughts of such authentically speculative content and of such a resolutely modern strike emerge intertwined, that one instantly thinks of the great systems of modern philosophical idealism, Leibniz or Hegel.
>
> Cassirer [28, p. 94]

Alongside passages that make man the sole repository of rational knowledge, the "place of all Reasons", we find numerological analyses of Pythagorean type:

> For, just as in the world the substances of things subsist, in man reside their rational reflections, their brilliance, their forms and true notions. If it is true that the world is indeed everything, it doesn't know anything. If it is true that Man is little, so to speak nothing, he knows all things. And just as the former is great in substance, so the latter is great in knowledge [...].
>
> As for artificial Man, that is to say, the specific human form engendered by art, it is a kind of emanation of primitive man, his Wisdom, his fruit and his end. By this acquired character, he who was by nature nothing more than Man, is now called twice Man by the benefit, by the superabundant advantage of art. [...] From these (two) extremes, there exists a certain concord and coincidence, a kind of love, peace, a kind of link, an intermediary, which is the product, the union, the fruit, the emanation of these extremes. Thus, the monad and the dyad, joined to each other, engender, produce the triad, which is their copula, their union and concord. That is why Wisdom is a certain threefold way of grasping Man, a

---

[6]It is common to write with a capital letter the metaphysical concepts and entities corresponding to common notions (Being, One, Idea...). This practice is shared by theology, Pythagorism, Platonism...: it is in reference to it that the use of capital letters throughout this book has to be understood, a use which, moreover, has been used with parsimony and only in contexts where it seemed necessary.

trinity, a humanity, a triad of Man. For the trinity is a rival of total perfection, since without trinity it is impossible to discover the slightest perfection.[7]

Bovelles [35, p. 354]

Numerology is not, in this passage of *Le Sage*, at the service of a scholastic obscurantism. Its use aims at supporting the idea, profoundly humanistic, of a necessary harmony between the corporeality and rationality of Man, an idea very present in French Humanism (Rabelais, Montaigne). It thus contributes to the foundation of a new conception of education, which rehabilitates the body and where the training of the mind must go hand in hand with physical exercise. The trinity, the number three, as a synthesis of the monad and the dyad, of the one and the two, serves here, by analogy, as a "mathematical model" for the necessary synthesis of the spiritual and the material.

It is far from certain that later thinkers (Spinoza (1632–1677), in his *Ethics*, for example), always made better philosophical use of mathematical thought. Caution therefore: the Renaissance ideal of rationality and equilibrium may well have, in all its incompleteness, naivety and theoretical and rhetorical imperfections, a depth too often neglected![8]

2.2 Presocratic Philosophers and the Possibility of Scientific Knowledge

If our knowledge of original Pythagorism is too fragmentary to lead to satisfactory conclusions, the way in which its idea of number was enriched by the contribution of later presocratic philosophers is easier to identify, thanks in particular to the accounts given by Plato (427–347 BC) and Aristotle. The influences of the presocratics on the philosophy of arithmetic being multiple and complex, it is necessary to have some preliminary reference points to understand them, as well as the role that arithmetic and mathematics played in the later edification of Aristotelian metaphysics.

It was Heraclitus (c. 550–480 BC), known as the "obscure", who seems to have first posed the problem of the conditions of possibility of theoretical knowledge. The world is a permanent flow, and the very stability of surrounding things is deceptive, subject as they are to becoming. According to three classical fragments: "We bathe and we do not bathe in the same river"; "We go down and we do not go down in the same river; we are and we are not"; "One cannot go down the same river twice". If the last fragment is the best known, the movement of dialectical opposition of

---

[7]Ch. de Bovelles alludes here to Pythagorean combinatorics, which makes the number 6, divisible by 3, the number of the accomplished perfection since 6, equal to $1 + 2 + 3$ and to the product $1 \times 2 \times 3$, operates the synthesis of the monad (1), the dyad (2) and the triad (3).

[8]It is to Thierry Gontier that I owe the discovery of the depth and richness of Renaissance thought. I refer the reader here to his works, which revisit classical Humanism and redraw its contours.

the first two seems more in line with the Heraclitean style of bringing opposites together. Above all, they bring to light the problem of Being, since, according to Heraclitus, "all sensible things are in a perpetual flow and cannot be the object of science" (Aristotle [6, A 6 987a]).

After Heraclitus, in order to restore the possibility of scientific knowledge, philosophers will have to relearn how to think about the permanent elements of this flow, for they are the ones that structure our language and our thinking. Indeed, it is necessary that Being—that which "is"—escapes in some way from the instability of becoming so that knowledge is not a mere illusion. It is perhaps this permanence that is already indicated by the fragment "We bathe and we do not bathe in the same river", together with the problematic union of the flowing and the permanent.

If knowledge seems to be subordinated to the moving character of its determinations, it is therefore necessary to assure solid foundations to science, which give back to Being that stability which is manifested in discourse. Platonism answered Heraclitus with the theory of Ideas,[9] suprasensible entities that escape the indeterminacy and the becoming that affect objects of senses. Aristotle, for his part, systematized the foundations of the theory of scientific knowledge with the theory of "Being as Being": the so-called First Philosophy, since it is the condition of possibility of all thought, or Metaphysics, from the title attributed by posterity to the Aristotelian work on the subject.

We must understand simply this formula "theory of Being as Being", made obscure by the weight of the philosophical tradition: if things "are", it is because our thought, our language declares them "to be". Now, this fact of being is structured. Being is said in different modes and is understood in various senses: substance (Socrates is a man), accident (what happens, but not always necessarily (Aristotle [6, K 8 1065a]): the musician who studies grammar is a grammarian by accident), potentiality (the bronze is potentially a statue (Aristotle [6, K 9 1065b])). In both Plato and Aristotle, numbers and arithmetic play an important role in this theoretical attempt to restore the possibility of science. Mathematics thus participates de facto and in various ways in Metaphysics, this theory which, let us insist, is a theory of knowledge and of its conditions of possibility against Heraclitean relativism.

2.3   Parmenides, the One, and the Birth of Metaphysics

Following the testimony of Plato [95], it is Parmenides[10] who would have been at the origin of the metaphysical questioning, of the formal theory of Being and of the first development of the themes that would link the latter to the philosophy of number. In order to be able to produce the fixed, permanent terms that knowledge

---

[9]See Chap. 3 of this book.

[10]Active in the first half of the fifth century BC. His dates of birth and death are not precisely known.

needs, it is necessary to be able to grasp signifying units, against the phenomenal background of a world that is in permanent becoming. The Heraclitean river is a good example of this: fixed as a conceptual referent, in perpetual motion as a thing of nature. The One is the constituent principle of these signifying units, since to think of an object, a thing, and to name it is to group and unify its different moments: its positions, its temporal becoming, its possible modifications of form. The river thus escapes indeterminacy as a unitary synthesis of its different moments—in flood, in such and such a bed... Hence the fundamental thesis of Parmenides: the One is none other than the Being. The thesis proved to be essential for the later philosophy of arithmetic, but is philosophically complex and difficult. Plato endeavoured to deploy all its complexity by developing the aporias to which it leads. We will come back to this.

In any case, the die is cast: after Parmenides the theory of numbers, through the theme of the One, becomes decisive for philosophical thought and the theory of scientific knowledge. Numbers, quantities, magnitudes, appear in this context of the birth of Metaphysics through one of the fundamental oppositions that structure Being: that of the One to the multiple or divisible (Aristotle [7, 13 1020a]). The slow decline of philosophy after Aristotle and the fourth century BC, its dispersion in the confused meanders of the Middle Ages, but also the great conceptual confusion that surrounds contemporary philosophy and its teaching of the classics, tend to obscure the original clarity of the Parmenidean equation: no consciousness, no thought that is not grasped as a unit, a gathering of different moments, whether it be under the figure of the spatio-temporal object torn from the flow of becoming or under the lasting figure of the concept.

The introduction of the theme of the One and its relation to Being by Parmenides would lead to various attempts to synthesize the Pythagorean mysticism of numbers with the guiding themes of Metaphysics. A particularly significant example is given by the *Treatise on Numbers* of Plotinus [97, Ennead VI-6],[11] a neo-Platonist of the third century AD. If Plotinus' thought marks, from the point of view of the philosophy of arithmetic, a clear regression with respect to classical Greek philosophy in its relative conceptual confusion, it bears witness to a form of spirituality which, although it is today marginalized, has nonetheless left its mark on a whole philosophical tradition. In methodological terms, leaving aside its chronological posterity, the reading of Plotinus will serve, for us, as an introduction to the major theoretical problems arising from the Platonic and Aristotelian philosophies of numbers, which will be discussed later.

---

[11]Translations of Plotinus will follow mostly [96].

2.4   Platonic Ideal Numbers

One of the fundamental distinctions established by the Platonic school and Neo-Platonism is that of Ideal[12] or essential numbers and counting numbers, which largely reflects the distinction of the Parmenidean One with the numerical "one" or, in more concrete terms, the distinction between the "metaphysical one", which is that of the spatio-temporal unity of an object or the intellectual unity of a concept, and the "one" of enumeration and calculation. Since the One is at the foundation of Being as one is at the foundation of number, and "since Being came into existence from the One, it must be number as this supreme One is one [...]. This number is the essential number."[13] (Plotinus [96, 9])

The relationships of these essential or Ideal numbers to counting numbers are elucidated later:

> We may be told: these numbers, which we call first and true numbers, where do you want to put them and in what kind of beings? [...] Not only do you say that these numbers belong to the first among beings, but you also say that in addition to these numbers there are other numbers, the counting numbers [...]. You owe it to yourself to clarify all these things for us.
>
> Plotinus [96, 16]

Various consequences follow from this with regard to the nature of numbers, ideal objects independent of thought, to which they would be anterior, or creations of the human mind confronted with sensitive phenomena:

> Plato, of course, when he says that men came to the notion of number by the difference of days and nights, when he relates its thought to the diversity of things, perhaps means that it is the things numbered that first produce the number by their diversity, that it is formed in the path of the soul as it pursues one thing after another, and that it occurs in the very moment when the soul numbers. [...]
>
> But yet, when he says that the number too is in the essence, he means quite the opposite, that the number has, of itself, a certain existence and that it does not come to existence in the soul that numbers, but that the soul awakens in itself, on the occasion of the difference observed in sensible things, the notion of number.[14]
>
> Plotinus [96, 4]

This Plotinian variant of the Platonic theory of reminiscence is also based on an argument of anteriority: in order to state a numerical property—whether ideal or phenomenal—the mind must already be in possession of the idea of number, which therefore pre-exists its discursive use:

---

[12]On the Platonic theory of Ideal numbers, see Aristotle [6, Metaphysics A].

[13]Armstrong [97] translates "Since, because Being came into existence from the One, and that One was one, Being must also in this way be number [...]. And this is substantial number."

[14]Armstrong [97] translates the last part: "But then, when Plato says 'in the true number', and speaks of the number in substance, he will, on the other hand, be saying that number has an existence from itself and does not have its existence in the numbering soul but the soul arouses in itself from the difference in sensible things the idea of number."

> If he who counts says that things are ten thousand, he does not say this because things
> would have shown him that they can be said to be "ten thousand" in the same way that
> they show their colour; he says this because discursive thought asserts that they have this
> quantity [...]. How can he say it then? Because he knows how to count, and he knows how
> to count if he knows the number; and he knows the number only if the number exists.

In fact, Plotinus distinguishes between ideal numbers and counting numbers, essential numbers and phenomenal numbers, according to their conceptual use. When the number appears at the principle of an essence, it is essential, where the number that measures quantity is simply counting:

> When you say that man in himself is a number, for example a Dyad,[15] living and endowed
> with reason, this expression does not offer a simple meaning. No doubt, as you pass from
> one thing tô another [from "living" to "endowed with reason"] and you count, you produce
> a quantity, but, as there are two things and each is one, if each of these ones contributes to
> the completion of the essence and the unity is in each, you say another number, an essential
> number.

An essential number is therefore the one that manifests itself within the unity of an essence determination, as when, in Pythagoras, 4 is attributed to justice or virtue—an example explicitly taken up by Plotinus [96, 16].

The type of philosophy of number that appears in Plotinus may appear to be a shaky compromise between Pythagorism and Platonism. It undoubtedly reflects an intrinsic and never really elucidated difficulty of mathematics, which is due to its dual status as a theory about ideal entities and as an effective theory, capable of giving rise to statements about the world, be it the world of everyday life or the theoretical constructions of the natural sciences.

2.5   The Lasting Influence of Neo-Platonism

It is in the nature of the issues surrounding number that one is constantly being taken back to the beginning. An example of the resurgence of neo-platonic themes in contemporary thought can be found in the book *Triangle of Thoughts* [31], which brings together a series of fairly informal discussions between three leading mathematicians: Alain Connes, André Lichnerowicz and Marcel-Paul Schützenberger. The book is a free-form discussion, and is not intended to be a treatise on mathematical thinking. This is perhaps its most interesting aspect: each contributor reveals his philosophy in a simple and direct way, which makes it easy to identify, in each intervention, the springs that underlie it. The three interlocutors, although all three mathematicians, have scientific personalities and epistemological aims that are markedly different and which, for various reasons and in spite of their modernity, are linked in several aspects, for the first two of them, to the Neo-Platonic and Pythagorean traditions.

---

[15]The Dyad, seen as an essential number, is to two what the One is to one.

Before going further, in the following chapters, into the classical and modern philosophy of numbers, it will not be uninteresting to see how mathematicians today spontaneously take a stand on the relationship between mathematics and ontology;[16] a quick analysis of this book will be the pretext for this. This digression will also be an opportunity to situate the debates on the nature of numbers and mathematical ontology in the field of current mathematical thinking. It will emerge that, contrary to appearances, a rereading of the classics, even when authors as "metaphysical" as Plotinus are involved, is not without consequences for the understanding of the moderns. Beyond *Triangle of Thoughts*, it seems that, in a very general way, the late commentators of Plato and Aristotle have deeply marked the collective unconscious, probably because of the theoretical influence, too often undervalued, of medieval theology and then of the Renaissance on the development of modern thought.

The thesis defended by Connes is the most immediately correlated with the Plotinian theses. Although supported by twentieth-century mathematics, it refers directly to the type of debates that marked neo-Platonism:

> I maintain that mathematics has an object, just as real as that of the sciences such as geology, particle physics, biochemistry, cosmology, etc., but which is not material, and is not localized either in space or time. It does, however, have an existence just as firm as external reality, and mathematicians come up against it in much the same way as one comes up against a material object in external reality. This reality of which I speak, because it is not locatable either in space or in time, gives, when one has the chance to reveal a tiny part of it, a sensation of extraordinary enjoyment through the feeling of timelessness that emanates from it.
>
> Connes et al. [31, p. 39]

Plotinus does not express himself differently, including in the idea of an almost mystical ecstasy that would be linked to contact with the "intelligible substance". For obvious reasons of thematic priorities, we have not insisted on the mysticism and the theological component of Plotinian thought, but they are inseparable from his conception of mathematics and condition it to a great extent.

The choice made by Connes, who adopts a language and terms close to mysticism in his description of the happiness of mathematical discovery, admits several interpretations. It can be understood simply as a reprise, more or less deliberate, of neo-platonic themes that contribute to structuring our collective unconscious. It can also be understood as a commitment to think about the springs of intellectual jouissance, as if it were intimately linked to profound mechanisms of thought on which Pythagorean or Platonic discourses sometimes manage to express themselves in a more adequate way than modern positivism.

Other features of the theses defended by Connes continue to bring him closer to neo-Platonism. Thus, mathematical "reality" would always infinitely exceed what can be apprehended from it, an idea that is quite close to the Plotinian opposition between the infinite perfection of the intelligible and a certain incompleteness of

---

[16]Ontology is the philosophical name for the theory of being (what is: why and how things, beings exist). In a mathematical context, ontology confronts the problem of the nature and existence of "mathematical objects" and "mathematical ideas".

the "world here below". Connes also justifies his epistemological theses with the help of one of the great advances of mathematical logic in the twentieth century: Gödel's theorems. There is something paradoxical about the resurgence of old metaphysical themes in the name of recent and relatively technical mathematical discoveries. Of course, Gödel's theorems (to which we will return in more detail later) have a rather special status in twentieth-century epistemology. They are at the origin of an abundant literature in the field of human sciences, of a content that is often debatable, Gödel (1906–1978) regularly serving as a scientific guarantee for these "intellectual impostures" that have been denounced by Alan Sokal and Jean Bricmont [106]. These theorems lead us once again back to numbers, their statement (and their proof) relating to properties of elementary arithmetic. In two words as a first introduction, there are undecidable statements in formalized arithmetic:[17] one can neither deduce their validity nor their falsity from the axioms. In some cases, however, it is possible to establish that these statements are "true", either by their very construction (in this case it is, broadly speaking, a semantic type of truth: their validity results from the very meaning we give to these statements), or by using methods that go beyond the framework of classical arithmetic.

Surprisingly enough, although faithful to the epistemological theses developed by Gödel himself, Connes does not react to these statements in the manner of logicians, who see in Gödel's results a formidable source of theoretical problems posed to mathematicians and logicians as to the very nature of the idea of proof or truth. He resolutely chooses, and this is what will interest us here, an ontological interpretation: "The distinction between truth and provability is one of Gödel's essential results. It is not the problem that is inexhaustible, it is reality itself." (Connes et al. [31, p. 58]) Connes is an emblematic mathematician of the end of the twentieth century and the beginning of the twenty-first. His work is profound and fertile, and it is with good reason that he thus puts back on the agenda, in a neo-Platonic spirit, the great problems of the theory of mathematical knowledge about the nature of objects and mathematical idealities.

2.6 Modern Structuralism and Pythagorism

A. Lichnerowicz, the second author of *Triangle of Thoughts*, defends a radically different point of view, but one that is also naturally linked to the theses of antiquity on numbers. He takes a so-called "structuralist" view of mathematical activity, which leads him to oppose any form of metaphysical approach to mathematical content:

> We have learned that if we do mathematics, it means that Being with a big B is put in brackets. This characterizes mathematics as a functioning of the mind in the sense

---

[17]That is to say, once a system of axioms has been fixed to manipulate the numbers and associated symbols.

that we simply want that, in this type of discourse, which is intended to be without misunderstanding, the "Being of things" is put in brackets. And this radically non-ontological discourse will be used by mathematics wherever the latter is practiced.

Connes et al. [31, p. 37]

This type of thinking is the heir to the axiomatic method, developed by Hilbert (1862–1943) at the turn of the twentieth century: a mathematical theory is given now by a collection of axioms and deductive rules. The statements, results and theorems of the theory are then deduced mechanically from the axioms, without any need at any time to appeal to the "nature" of the underlying mathematical objects. The idea is very appealing and corresponds to deep phenomena in mathematical practice and history, but it is doubtful whether it is able to account for all dimensions of mathematical thinking, except to make it a pure game of the mind disconnected from all meaning and applicability. Thus, the experience of beauty or the joys of discovery, as described by Connes, hardly finds a place there, even though it constitutes one of the foundations and one of the reasons for mathematical activity—and for any passion for science.

To stay with the example of numbers (natural, rational, real, complex...), the architecture of the systems of axioms that make it possible to account for their construction and properties depends on decisions that are anything but spontaneous and innocent. This is indeed the weak point of the refusal of ontology: a visceral incapacity to confront real mathematical practice in all its dimensions, the one that Connes reports, where the mathematician is not dealing with empty symbolism but with objects that have meaning and very often have real-world interpretations—in physics, for example.

The perverse effects of structuralist thinking are moreover especially visible through its reception by the human sciences.[18] The guiding idea that has governed this reception is identical to that defended by Lichnerowicz: it is a question of putting the Being in brackets in order to arrive at a discourse without misunder-standings, a description stripped of all roots in concrete reality of the mathematical structures underlying a given phenomenon. The errors to which the method leads always follow the same drift: an object of study is led back to a formal explanatory scheme (this is the process of axiomatization). This explanatory scheme will then be interpreted as a formal cause—it passes from the status of a descriptive model to that of a structural explanation. Finally, in a third and final movement, the formal cause becomes an effective cause, the real vanishes in front of the underlying "structure".

Paradoxically, structuralism, the "modern" theory par excellence, ends up join-ing, in its internal logic, Pythagorism. The ultimate consequence of structural thinking when it is thus applied to reality is indeed a fetishization of mathematical concepts, and the temptation to think of the essence of the reality as the underlying mathematical structure. The "phallus equals root of -1" of Lacanian psychoanal-ysis[19] as well as the adoption by Piaget (1896–1980) of the theory of "mother

---

[18] With the works of Piaget, Lévi-Strauss, Lacan....

[19] Which aims to codify through mathematics the repression of sexuality.

structures" (ordered, topological...),[20] two theses characteristic of the adoption and adaptation of the theses of mathematical structuralism by the human sciences in France, in the 1960s, are, in their ultimate principle, only sophisticated variants of the reconduction of the real to formal schemes that is characteristic of Pythagorism. We would thus be dealing with a form of renewed Pythagorism, where structures, the formal objects of modern mathematics, replace, *mutatis mutandis*, the numbers of the first Pythagoreans.

Professional mathematicians such as Lichnerowicz were careful not to push the logic of their theory to that point, but the risks of deviations were almost inevitable because of the normative vocation of the endeavour and its "non-ontological" orientation, which prevented it from thinking about the relationship of mathematics to reality in all its complexity.

The fundamental trait of Pythagorism: to make the mathematical model and, by extension, any formal structure, the essence of the corresponding real, is thus found in modern societies and science, in forms that are increasingly subtle, and sometimes brutal. This process leads to confer autonomy to formalized mathematics with respect to its applications and its intuitive foundations. However, mathematical models, wherever they are applied, are always meaningful relative to given situations, and are, for example, sensitive to phenomena of scale (as in physics) or to simplifying hypotheses (as in economics). When the methodological importance thereof is minimized or lost sight of, the very meaning of mathematization is distorted. To claim that the "non-ontological" or "non-philosophical" character of twentieth-century mathematics may have contributed to a degeneration of the notion of modelling, particularly in the human, economic and social sciences, may therefore not be without legitimacy.

In mathematics proper, the fetishism of the formal too often leads to rigid teaching methods and limits the impetus for creative thinking. Even in the most abstract theories, recourse to intuition, a look back at the scientific universe in which these theories were built, the possibility of imagining, of "playing" with concepts by operating on the transformations that one allows oneself to make them undergo, are all necessary moments in the construction of new theories that invalidate epistemologically and ontologically neutral discourses. Post-Heraclitean Greek thought had already taught us this: to put the "Being of things" in brackets is to put the essence, the principle of theoretical knowledge in brackets. In this, Connes is right against Lichnerowicz.

---

[20]With important consequences in educational terms. Piaget's educational theory stems from the thesis of "mother structures" and codifies the emergence of mathematical thinking in children from these structures.

2.7 Aristotle on Numbers and Mathematics

Neo-Platonic idealism and the temptation to absolutize mathematical content, to make it the norm of the real, are permanently inscribed in our conceptual heritage. Most of the most debatable Pythagorean or Platonic thought patterns were nevertheless refuted by Aristotle in his *Metaphysics*, especially in Book M, but these refutations did not have the posterity and success they deserved. One of these refutations concerns the mode of ideality of numbers and mathematical notions, and shows how certain epistemological debates that are still relevant today are rooted in the invention of Metaphysics. In addition to its relevance for the philosophy of arithmetic, it shows that another path is possible for mathematical philosophy, too frequently trapped, as we have just seen, between a naive and vaguely mystical idealism and a harsh formalism, forgetting reality and hypostasizing structures to the detriment of their rooting in thought and reality.

The Platonic theory of reminiscence rests on an argument of anteriority: no thought is possible without a pre-existence of the concept, the essence, the thing to which the statement relates. In terms more in keeping with the spirit of the Parmenidean tradition: no thought, no discourse is possible without a permanence of the objects of discourse that transcends the temporal, flowing character of thought. Or, in more modern terms: mathematical objects, first and foremost numbers, pre-exist their discovery. All this is not entirely false according to Aristotle: it is solely necessary to agree on the terms and not to think of this anteriority in a substantial way—which is what the neo-Platonists as well as the modern Platonists do.

The mode of existence of mathematical objects, just like that of ideas, has the peculiarity that if one can "in all truth grant being to mathematical things, and with the characters that mathematicians assign to them" (Aristotle [6, M 3 1077b]), this mode of being is non-substantial: the mathematical circle is the abstraction of a form, not a Being in its own right, nor even the substance of an actual being, and it is according to different modalities of knowledge that we will say that "Socrates is a man" and that "this figure is a circle". The formal analogy between the two statements is misleading, conceals a difference in nature, and leads to the attribution to the objects of mathematics of attributes that are specific to everyday objects and beings. This leads to a whole set of aporias and "badly posed problems" that often bog down debates on mathematics.

With respect to the anteriority of the mathematical object to statements about it, on the basis of the theory of reminiscence and the thesis that mathematical activity is an activity of discovery rather than creation, care must be taken to distinguish logical anteriority from substantial anteriority. The objects of the surrounding world are anterior to the discourse we carry about them: they "are there", and it is this constant presence that guarantees, beyond the minor transformations they may undergo (the Heraclitean river), the validity of the discourse. Because of their substantial anteriority, they "pre-exist" the discourse we may carry about them. Ideal objects are also anterior to discourse, in the sense that to say that "there are five chairs" presupposes the existence and a certain permanence in our mind and in the collective consciousness of the number 5. This anteriority is logical: we must know the number 5 for the statement "there are five chairs" to have any meaning. This does not mean, however, that the number 5 has any "eternal substance" that would guarantee this anteriority in the order of discourse, since, unlike the objects of the surrounding world whose presence and use guarantee their existence, there is nothing allowing us to affirm that the number 5 exists outside the discursive and cognitive universe in which it is constituted.

Aristotle's position, which will implicitly be the one adopted in this book, is ultimately intermediate between those of Connes and Lichnerowicz: a reflection of ontological type on mathematical thought is necessary, leading us to think about the nature of mathematical objects in terms of the permanence of forms, but cannot be limited to neo-Platonic realism. Mathematical entities are therefore idealities and not objects in the proper sense, but this status of ideality is sufficient for the mathematician to proceed in the majority of cases with mathematical things "as if" the abstract circle or the number 5 were objects: we can "in truth grant being to mathematical things".

Chapter 3
The One and the Multiple

In the previous chapter we saw the birth of theoretical thought and the emergence, with Parmenides, of the theme of unity and the One in the theory of knowledge. Together with this theme goes the theme of multiplicity, in terms of a dialectic that is the object, in philosophy, of mereology, the theory of whole and parts. Plato's dialogue, the *Parmenides*, tackles this dialectical opposition of the One and the Multiple head on. It is one of the fundamental texts in the history of philosophy, but also one of the most difficult. The necessity of idealism is resolutely affirmed as the only one capable of founding the possibility of thought against relativism and the Heraclitean flow:

> If one does not admit that there are forms of beings, and refuses to assign a form to each of them, one will no longer know where to turn one's thought. Indeed, doing so, one does not want an idea that is always identical to exist for each of the beings. Thus, one completely destroys the possibility of discussion.[1]
>
> Plato [94, 135]

The objective of the *Parmenides* is in fact twofold: to found methodologically the study of the forms of beings and thus establish the possibility of theoretical knowledge; to study according to these principles the concept of the One as put by Parmenides at the foundation of thought and Being.

The influences of the *Parmenides* on the idea of number were multiple. The reading is all the more delicate since Plato does not expose a constituted theory, but rather the difficulties of philosophical research, constantly confronted in its movement with new aporias, new contradictions. His conclusion, which summarizes

---

[1] Allen translates [95, 135] "if one will not allow that there are characters of things that are, and refuses to distinguish as something a character of each single thing, he will not even have anything to which to turn his mind, since he will not allow that there is a characteristic, ever the same, of each of the things that are; and so he will utterly destroy the power and significance of thought and discourse."

© The Editor(s) (if applicable) and The Author(s), under exclusive license
to Springer Nature Switzerland AG 2020
F. Patras, *The Essence of Numbers*, Lecture Notes in Mathematics 2278,
https://doi.org/10.1007/978-3-030-56700-2_3

these aporias, may have seemed to some commentators to reflect an irony of Plato with respect to Parmenides. It is more likely that it illustrates simply the complexity of the ontological status of the One:

> Let us say it, then, and add that, to what it seems, whether the one exists or does not exist, it and the other things are and are not, appear to be and do not appear to be absolutely everything, relatively to themselves and to each other.
>
> Plato [94, 166]

Let us develop, as an indication of the Platonic way of thinking, one of these aporias. Let us assume that the One exists. It cannot be multiple without contradictions. The One cannot therefore be a compound or have parts. Yet, the Being is distributed among all the many beings that exist and, by virtue of Parmenidean thought, the One is never absent from the Being, nor the Being from the One. The One is thus shared in the same way as the Being, and is therefore multiple.

This aporia has had various ramifications, among others, in medieval Muslim and Christian theology. The arguments developed around the Trinitarian doctrine (father/son/healthy spirit) thus largely obey the logic of the terms of the Parmenidean aporia. Let's quickly indicate how the aporia can begin to be resolved. It is first of all in different senses that the One is said to be multiple in the first part of the reasoning and multiple in the second. The number one is indecomposable for arithmetic: we will never find non-zero positive integers p and q such that $1 = p + q$. In this sense, the One is not multiple. The equality $1 = 2 \times \frac{1}{2}$, which would suggest that the One is divisible, is based on a scientific logic that goes beyond the epistemological context that interests us: that of natural numbers. It only makes sense in a more general context (geometric or symbolic, for example). It should be noted in passing that the epistemological status of fractions and the determination of their exact relationship to the concept of number were discussed and were still the subject of polemics at the beginning of the twentieth century. This is amply demonstrated by the conclusion (Schluss, chap. V) of the *Foundations of Arithmetic* [50] by G. Frege (1848–1925) or volume XXI [74] of the complete works of Husserl (1858–1938). We shall return to these questions later.

This duality of points of view (arithmetic indecomposability of one; algebraic decomposability) is essential to understand the springs of the philosophy of number and the meaning of current mathematics. The same object (an integer, a sphere, a triangle...) is indeed likely to "live" in multiple mathematical universes, and its meaning and its very properties depend each time on this universe. One, indecomposable as an integer, is no longer such when seen as a rational number. The sum of the angles of a triangle is π in Euclidean geometry, but this property ceases to be true in hyperbolic geometry. In contrast to the geometric sphere, the topological sphere, an abstract gluing of two discs along their edges, has no intrinsic metric properties. The examples could be varied infinitely. The mathematical object, apart from being an object only in an improper sense, therefore does not have an absolute existence, but always relates to a given use, context and theory.[2]

To come back to the Parmenidean reasoning, which identifies the One and the Being, it rests on principles and on a logic that are not algebraic but rather set-theoretic, and it is in this set-theoretic context that we must seek to analyze the aporia. To say that Being is One is to say that every "thing", every quiddity, can only be thought of as a unit—even if it is multiple, like the human body, composed of parts. To say that the One has the multiplicity of Being is to state in a different way this possibility of subsuming any quiddity under the conceptual regime of the One—Parmenides' fundamental intuition. But these quiddities are not parts of the One, in the sense that their union would compose the One: a concept is not the sum of its instantiations. It is therefore improper to claim that the One is multiple, because the One is then treated as a totality in a context where it functions rather as a concept. As we shall see, it was only at the end of the nineteenth century, in Frege's research on the foundations of arithmetic, that the tools for thinking about these distinctions were put in place satisfactorily, in a properly mathematical way.

3.1 From Plato to Frege

One of the lessons of the history of the concept of number is that, if one tries to read the texts with a mind open to conceptual problems, deep similarities emerge between the questions and the answers given to them from one era to another. These similarities are much easier to detect than in other philosophical or epistemological fields, since the precise content that can often be associated with intuitions or arguments about numbers makes comparisons between authors and eras much easier.

In the case of the *Parmenides*, we have seen that Plato seeks to promote the thesis of Ideas, but also to illustrate its methodological difficulties through the exemplary case of the One. A similar approach can be found in a text by Frege from the years 1891/1892, "On the concept of number" [54]. In a distant echo to Parmenides' desire to legitimize theoretical knowledge, Frege seeks to illustrate the difficulties of arriving at a scientific definition of fundamental mathematical concepts. Frege's ambition is quite different from Plato's: he wants to arrive at a purely analytical[3] theory of number. In spite of this difference in perspective, Fregean theses have in common with Plato's the claim that mathematical objects are objects in their own right.

---

[2] One should not conclude from this, too hastily, the relativism of mathematical knowledge. Indeed, within a given theory, the object keeps a fixed and well-defined identity. What we want to draw attention to is the fact that the multiplicity of occurrences of the same mathematical term in different contexts can naturally induce contradictions, which can be resolved if we place each time the use of the term in the appropriate conceptual environment.

[3] In a logical and philosophical sense.

Frege attacks a definition of number that is fairly classical in its conception and that is representative of a naturalism tinged with psychology that was frequent until the end of the nineteenth century. The definition is proposed for example by Otto Biermann in his *Treatise on Analytical Functions* [12]:

> The concept of number can be defined as the idea of a plurality composed of things of the same species. When we use the term 'one' for each of the elements of the same species, counting the elements or units of the whole consists in assigning the new terms 2, 3, etc., to one and one – and one, etc. The number is the idea of the group of elements designated by these terms.

The Fregean analysis focuses on the springs of Biermann's definition. Thus, the word "idea" does indeed have a meaning in ordinary language, and also in philosophical language, but its use by Biermann is a mere delusion for lack of adequate precision. Frege's style is very direct and only becomes complex when what is being thought of requires it because, in his own words: "There is no better place to hide the most childish confusions than in the most seemingly sophisticated terminology", a classic flaw of academic thought. So he chooses simple examples: Suppose there is a lion at the side of a lying lioness. Together they make a group of things of the same kind. Would we then say that the number 2 is the idea of this group of lions? Beyond the psychologically unconvincing nature of the assertion, problems arise immediately because it is impossible to dissociate the idea from its material roots (this group of lions, whose idea cannot be identified with that of a group made up of a tiger and a tigress). The word "idea" is therefore insufficient to describe the counting process. To proceed now by associating the term "one" with the lion, then with the lioness, to conclude that two is the idea of this "one" and this "one" put together leads to other aporias, because the one would first have to be defined in a different way than as a simple marker. In short, a rigorous conceptual analysis to define unit and number is required, and Biermann's "definition" is at best a plausible description of one of the possible uses of numbering; in no case can it claim scientificity and normativity.

3.2   From Frege to Wittgenstein

It may seem ridiculous to approach the idea of numbers through such elementary considerations, but this is a mistake of perspective. In fact, modern mathematical philosophy has paid too much attention to abstract phenomena (axiomatic theories, logical paradoxes...) to the detriment of the study of the roots of mathematical thought in certain fundamental phenomena (the experience of the world, abstract processes, language). The understanding of natural numbers, if it ultimately requires a great deal of skill, as evidenced by the long list of mathematicians and philosophers who have dedicated an important part of their work to it, is ultimately based on the analysis of fairly simple thought mechanisms. Here, the difficulty lies not in the complexity of the underlying mathematics, but in its articulation with the general principles of theoretical knowledge. From this point of view, we must today relearn to think about certain things, such as the extraordinary flexibility of ordinary language which, in its mathematical use, conceals behind formulas with an elementary structure (the concept... is the idea of...) complex logical constructions and mechanisms (synthesis, inference, abstraction).

Frege's work goes in this direction, but it is not the only one. The thought of Wittgenstein (1889–1951), whatever the legitimate reticence that a mathematician may feel towards some of his theses, has thus undoubtedly something to contribute, perhaps more in terms of approach (a certain freedom of tone, an undeniable ability to think ahead of ordinary conceptual schemes) than of content. In contrast to Frege's thought, strongly influenced by the classical philosophy of knowledge, especially Kantism, Wittgenstein's thought is radically different from classical philosophies, hence its interest but also its limits. For Wittgenstein, knowledge is less about concepts than about linguistic practices and usages. If it is therefore possible, albeit problematic, to associate meaning with mathematical statements, it will be through the use that can be made of them in non-mathematical statements about "states of fact". As long as they remain internal to mathematics, the same statements only have a "grammatical" status: rather than producing or manifesting knowledge, they just tell us how "mathematical words" have to be used.

This conception is quite strongly counter-intuitive since it not only challenges a naive conception of the "mathematical object", but seeks to think about mathematics without any reference to the theoretical and conceptual practice that goes with it. Hence, for a professional mathematician, a feeling of uneasiness when reading texts as different as the *Tractatus logico-philosophicus* [112] or the *Lectures on the Foundations of Mathematics* [113]. The dominant impression is that the theories developed there are entirely incapable of giving an account of the real functioning of mathematics. In spite of all these limitations, partly pointed out by Wittgenstein himself, his reflection on mathematics allows us to better understand, as far as numbers are concerned, the theoretical stakes of "naive" analyses, such as those carried out by Frege in "On the concept of number".

The key observation is that one of the reasons for Biermann's difficulty in moving from the consideration of a group of concrete objects to a numerical statement is that there is a difference in nature between everyday experience and the discursive phenomena at work in "pure" mathematics:

> "20 apples + 30 apples = 50 apples" may not be a proposition about apples. Whether it is one depends on its use. It *may* be a proposition of arithmetic – and in this case we would call it a proposition about numbers.
>
> Wittgenstein [113]

The challenge for Wittgenstein will therefore be to "show the difference between mathematical propositions and experimental propositions that look exactly the same". To a certain extent, it is impossible to dissociate statements in ordinary language from their use and their mode of formation, since the same statement can have multiple statuses: the phrase "It must be a Tintoretto!" will not have the same meaning and function depending on whether it is pronounced by a tourist visiting Venice or whether it appears in an expert's report!

We must see here the origin of the ambiguities in the definitions of number such as Biermann's: by claiming to define number as the idea of a plurality composed of things of the same kind, Biermann in fact presupposes his readers' empirical knowledge of the rules of use of numerical statements. We can see this with his characterization of the number 2 as the idea of a group made up of a lion and a lioness: anyone who does not know what a number is would have a hard time understanding what it is all about! In other words, one must already know what a number is and how it is used in ordinary language in order to understand Biermann's definition. This is what irritates Frege, who is trying to obtain an ex nihilo, purely logical definition of the number, which would not refer to an established practice or a phenomenal foundation. Wittgenstein's approach, in all its anti-idealism (a statement must be understood in the context of its empirical use), has the merit of significantly clarifying the terms of the debate.

3.3   Aristotle: The Whole and the Parts

In his critique of Biermann's views, Frege then develops another example, fundamental to the understanding of the philosophical problem of numbers and intimately related to the Parmenidean aporia reported at the beginning of this chapter. For this he considers the Laocoon group, the statue of the priest of Apollo and his two sons, who, according to Trojan legend, perished, suffocated by an enormous snake for having offended the god. In the statue, the bodies of Laocoon, his sons, and that of the snake are entangled and sometimes locally indistinguishable. So, if one asks what number should be associated with the Laocoon group, several answers are possible: one, if one considers the unity of the group, the idea of the statue seen as a whole—in accordance with the purpose of the statuary, which was to lead to a perception of the inextricability of the bodies. Four, if one tries to individualize the living beings represented—but this individualization could be made difficult by the artistic fusion of the bodies: it is supported by the art lover's prior knowledge of the scene, knowing to which legendary situation it refers and which assures him, for example, that there is only one snake. Finally, an indeterminate but very large number, if we consider that the common terms by which to discriminate the components of the group are the calcium carbonate molecules that make it up.

What can we conclude? And what does Frege want to suggest with such an example? No mathematical proposition is experimental. No more than any other mathematical idea, the one of number is material: it is not inscribed in reality as a property of things. A number statement, however simple it may be, presupposes abstract mechanisms and a conceptual regulation.[4] Most often this regulation is self-evident in the empirical uses of numbers and, for example, the question does not arise of the mode of individualization of objects/units when counting. This occurs for many reasons that are sometimes cultural (we evaluate a book by the number of pages rather than the number of printed characters, which would be more relevant, as all authors know), sometimes topological (the objects to be counted are most often physically separated, the case of the Laocoon being quite exceptional).

However, the question of the nature of the conceptual regulations at work in counting is raised. Without an answer to this question, any agreement on number could well be only formal: detached from the empirical and intuitive conditions of application of the concept to reality. This is one of the major problems of contemporary epistemology and didactics: mathematics can very well be understood and taught autonomously, without any reference to its effectiveness, to its ability to inform reality. The study of the ways in which young children learn numbering or of pathological cases (linked to the loss of skills following brain damage, for example) clearly show how difficult it is to stop there when, in cognitive terms, numbering is a complex and multifaceted phenomenon in which the formal and operational aspects (calculation techniques, etc.) are subordinate to more fundamental structures of thought, language and the brain.

With regard to the dialectic of the One and the Multiple, the Parmenidean aporias have historically found a provisional solution in the Aristotelian theory of the whole and the parts:

> A whole is that which contains things in such a way that they form a unit. This unity is of two kinds: either as the contained things each have a unity, or as the result of these things forming a whole. In the first case, the universal and what is said, in a general way, about the whole, is universal in so far as it embraces a multiplicity of beings. It is asserted of each of them, and all of them are one in the sense that each is unity: for example, man, horse, god are one, because they are all living beings. In the second case, the continuous, the limited is a whole, and unity results from several constituent parts.[5]
>
> Aristotle [6, Δ 26 1023b]

Unity is thus said in multiple senses, which will induce as many variations in the idea of number. First there is conceptual unity, whose archetype is the case of universals. It will be the principle of the modern extensional definition of number. A concept, like that of a living being, is a unit, a whole that embraces the objects that

---

[4] And in this, Frege's discourse goes much further than Wittgenstein's, who was largely incapable of taking into account the normative dimension of the concept beyond its empirical and discursive uses.

[5] Kirwan [7] translates "We call a whole both that of which no part is absent out of those of which we call it a whole naturally; and what contains its contents in such a manner that they are one thing, and this in two ways, either as each being one thing or as making up one thing. For what is universal and what is said to be as a whole, implying that it is a certain whole, is universal as containing several things, by being predicated of each of them and by their all—each one—being one thing; as for instance man, horse, god, because they are all animals. But what is continuous and limited [is a whole] when it is some one thing made up of more than one thing, especially when these are potential constituents of it but, if not, when they are actual."

depend on it (man, horse, god). The same applies to concepts that are not strictly speaking universals (the concept of the "lion here present", for example). In both cases, unity manifests itself at two levels: unity of the concept, unity of each of the objects falling under the concept.

The second case is significantly more complex, since unity (the whole) can be formed in different ways: topological (the indivisibility of the block of marble of the Laocoon), or functional: the unity of an achievement (that of the parts of a boat) or the bodily unity of natural beings. In all these cases, the whole is well grasped as a unit, but does not allow itself to be naturally decomposed as a multiplicity of units.

Finally, as for the number, which we would like to view as a whole itself made up of units, it requires a specific analysis, since a number is at first glance a unit made out of its components neither in the mode of universals or concepts, nor in the mode of a functional or topological unit. Aristotle thus introduces, in the study of the whole and the parts, a distinction which, without really resolving these aporias, will prove decisive in modern attempts to construct numbers:

> Moreover, [limited] quantities in which the position of the parties is indifferent are called a sum, and the others a whole [...]. Water, all liquids, and number are said to be only a sum, the word whole not applying to number or to water, except by extension.

What's that supposed to mean? Let's take the example of this book: its pages form a whole, which would be distorted and illegible if their order was arbitrarily modified, page 15 succeeding page 132, etc. However, for those who are only interested in the size of the book, in the number of its pages, this order is indifferent, and they can very well, in order to calculate this number, start by undoing the binding and mixing the pages: in this sense, the "sum" of the pages of the book is a less rigid concept than that of the "whole" of its ordered pages. From the point of view of modernity, the power of the Aristotelian analysis lies in part in its precise and mathematicizable content, since modern algebra often characterizes this type of process (the passage from the whole to the sum) as a "passage to the coinvariants".[6]

This idea, formulated in a more mathematically precise way, can be found in Kronecker (1823–1891) at the end of the nineteenth century. It is a very general mechanism whose formalization is not entirely immediate, but whose general idea is easy to understand. Let us suppose for example that transformations operate on an

---

[6]More specifically, the idea of switching to coinvariants generally refers to the action of a group, such as a group of substitutions or symmetries, and amounts to taking the quotient by this action. The shuffling of a deck of 52 cards is an example, the underlying group being the group of permutations of 52 elements. Many elementary notions, such as that of vector, are often implicitly defined as coinvariants under group actions, such as displacement. This notion of coinvariance, mathematically dual to that of invariance, was fundamental for the development of mathematics in the twentieth century, in algebra with group theory, but more generally in all mathematical disciplines. In geometry, the idea of invariance/coinvariance appeared as a constituent of the very idea of space, with the work of Felix Klein, at the end of the nineteenth century; algebraic topology studies the algebraic invariants/coinvariants of forms; in probability, the phenomena of temporal invariance are fundamental, etc.

object by moving some of its parts: shuffling of cards, continuous deformations of a rubber object, etc. The object is deformed, modified by these transformations, and yet its "nature" is preserved, the latter remains invariant under these transformations or deformations. Thus, if a book whose pages are shuffled ceases to be a book proper, if a chess game situation is distorted when the pieces are moved, a pack of cards[7] remains the same pack once the cards have been shuffled, in the same way that a tyre that is inflated remains the same tyre. In Aristotelian terminology, a book, a position in a chess game, a deck of cards are wholes. The pack of cards, a sum. Numbers belong to this last family, the order of the units which compose them being indifferent.[8]

To conclude on this theme of invariance, our activity of ordinary judgment proceeds by implicitly distinguishing sums and wholes, that is to say by distinguishing systems of objects for which a certain type of transformation is allowed that preserves, or does not preserve, their nature. We spontaneously distinguish two positions in chess games, but are indifferent to the way coins are mixed in our wallet, only their total value being important. The Heraclitean river, even flooded, always remains the same river. The problem of the object as it arises in Greek thought and mathematical philosophy is thus, in fact, intimately linked to the idea of transformation. Mathematics has only recently become aware of this, among other things through the idea of category.

Beyond the partition between sums and wholes, the Aristotelian distinction between unity of the concept[9] and functional unity (that of a boat, of the human body) is subject to infinite variations, which give the measure of the extension of the domain of formal ontology. The epistemology of numbers of René Thom (1923–2002) is particularly interesting from this point of view, since it can be partially inscribed in the lineage of the idea of functional unit in its topological and geometrical form. This makes it the support of an alternative construction of numbers. Thom's theses illustrate an important current in the philosophy of mathematics of the 1980s (catastrophe theory).

> Let's take an example that is dear to me [...]: arithmetic addition. Logicists have defined arithmetic in an inductive way, by means of the axiomatic definition of the successors of 1...
>
> The Peano system...
>
> Yeah, the Peano system. However, I think it is an artificial presentation.[10] The definition of addition, as I understand it, is the situation where you have the elementary catastrophe

---

[7]We distinguish here between a pack of cards (a set of cards, where the order doesn't matter, as when players look for a pack of cards to start playing) and a deck of cards (an ordered set of cards, for example a particular deck that is going to be distributed among the players).

[8]These notions of invariance by transformation and their links with group theory are well-illustrated by Hermann Weyl's essay, *Symmetry* [110].

[9]As already mentioned, two types of unity are actually attached to a concept: the unity of a totality, such as that of the universals [man], which embraces all their incarnations, and the unity that is proper to each object falling under the concept [Socrates].

[10]Thom refers here to the axiomatic-formal definition of number, which, it must be said, has never claimed to account for the meaning of the concept of number. To take a stand on the logicist

of one potential well being captured by another. In each potential well, there are objects that can be considered solid in the sense that they cannot mix. In the process of capture, the objects in one potential well fall into the other and all that remains is to count the number of objects that one has after capture. This is a bit like what children were taught in the past: to add the eggs in two baskets, you pour the contents of one into the other... The underlying process is continuous. In general, I believe that the only interesting mathematical structures with some legitimacy are those that have a natural realization in the continuum.

Thom [107, p. 146]

Thom is thus resolutely opposed to a discrete approach (in a topological sense) to units and to the concepts of number and number operations that go with them. These are the ones that mathematicians have historically preferred, favouring the unity of the concept to the detriment of the functional unit, and interpreting ultimately the subordination of an object or a being (Socrates) to a concept (man) in set-theoretic terms. The "Thomian" conception of number, on the other hand, is geometric and dynamic. It reflects the anchoring of the work of thought in temporality, space and movement.

From these different approaches, it should be remembered that number and unity are expressed in multiple senses, articulated in as many original experiments to which it would be possible to associate mathematical models and conceptual foundations of arithmetic.

3.4   The One, the Multiple, Man

Let us return for a moment, at the conclusion of this chapter, to Pythagorism, Neo-Platonism and their propensity to make abusive analogical use of mathematics. It is quite instructive to observe how the analysis of the One and the Multiple evolved. Because of the proximity of theme of the One with that of the Good in Plato and then in Plotinus and the Neo-Platonists, this analysis transformed itself into the one of ethical and theological problems, and this had, like Pythagorean numerology, a lasting effect on the collective unconscious.

It is all about the possibility of an analogical and metaphorical use of the notions of unity and multiplicity through a dynamic and geometrical conception of these notions. Thus, according to Plotinus:

Is it true that multiplicity amounts to taking a distance from the One, and infinity is a total distancing from the One, because it is a multiplicity impossible to enumerate? Is that also the reason why infinity is evil, and we ourselves are evil when we are a multiplicity? In fact, everything is a multiplicity if it is incapable of inclining towards itself, and if it flows and spreads out, it is a multiplicity. If it is totally deprived of the One in this flow, it becomes a multiplicity in which that which unites its parts to one another no longer exists; if, on the other hand, in the course of this flow, it becomes something stable, it becomes a magnitude. But what is fearsome about magnitude? For a conscious being, there would be something

definition (Frege, Russell) would have been more significant. On Peano's system and the logicist approach, see Chap. 12 of this book.

fearsome about it, for he would be conscious of taking distance and moving away from himself. For every being seeks not another but himself, and to go out of oneself is nothing but vanity or necessity; and every being exists more, not when it becomes multiple or great, but when it belongs to itself: it belongs to itself if it is inclined towards itself. As for the desire which tends towards the wrong way of being great, it is the fact of whoever ignores the true way of being great, and whoever strives not where he must but outside himself...

Plotinus [96, I,1]

"Being oneself" would thus correspond to a natural "inclination": in René Thom's vocabulary, the unity of the self is guaranteed by the existence of a well of potential. He who refuses to follow the slope of this well loses his unity, flows out of himself, scatters, gets lost. The moral connotation of this physico-mathematical model—since it is indeed one—is obvious, and has already become part of our vocabulary: it is necessary to know how to be oneself, not to disperse, true greatness is interior, and so on.

This permeability of all discursive thought, even in ethics, to mathematical images or analogies is striking. This is another of the reasons why mathematical education must teach the use of mathematical tools in the modelling of reality, so as to guarantee students a minimum distance from all discursive uses of mathematical models and analogies. As for the unity of the self, in order to have meaning and ethical implications, it must not be limited to the very Pascalian refusal of dispersion of the self and to the Plotinian refusal of hybris, but looking for it must go together with a quest for a functional unity. Rather than the passive unity of the determinations of being underlying the will to "be oneself", one should prefer the dynamic unity of a destiny that is constructed in action in the stoic manner.

Chapter 4
Mathematics and Reality

Numbers and, more generally, mathematical objects are not objects in the usual sense: they have no real and, a fortiori, no material existence in an improbable sky of Platonic Ideas. However, the idea of number does exist, algebra is there to testify, and it cannot be reduced to the empirical use of numbers in counting; hence the general problem of participation,[1] namely the relationship of mathematical ideals and concepts to their empirical use. This is a central difficulty for the whole philosophy of mathematics and the epistemology of the natural sciences, which therefore goes far beyond the framework of the philosophy of arithmetic, and for which there is as yet no truly satisfactory answer.

A concrete illustration of this problem is given by the attempts to justify the multiplication of quantities up to modern times. The Greeks were well aware of the underlying difficulties. Thus, if one multiplies two geometric measures (for example two lengths), what one obtains is a quantity that is not homogeneous to the multiplied quantities (in the example under consideration, one obtains an area, that is, a measure of surface). Modern algebra has taught us to overcome the reluctance to multiply inhomogeneous quantities—sometimes abusively: anyone who has taken physics courses knows that the first thing to do when verifying an identity or a result is to check that the quantities obtained have the correct physical dimension (that of an energy, a mass, etc.), because it often happens to beginner physicists that they get carried away by mathematical formalism and forget the physical meanings, thus obtaining, for example, as a measure of the earth-moon distance a quantity whose physical dimension is that of a duration or an acceleration.

---

[1] In the sense that, for example, the idea of number participates in our understanding of quantities, or that Euclidean geometry participates in our understanding of space—with notions such as straight lines, parallelism, right angles... Pythagoreans and Neoplatonists have a stronger definition of participation in the sense that, for them, mathematical entities, at the forefront of which are numbers, participate in the being of things and not only in their intellection.

© The Editor(s) (if applicable) and The Author(s), under exclusive license
to Springer Nature Switzerland AG 2020
F. Patras, *The Essence of Numbers*, Lecture Notes in Mathematics 2278,
https://doi.org/10.1007/978-3-030-56700-2_4

It is however edifying to note that these questions were still a source of embarrassment even after Descartes (1596–1650). Thus, in his *New Elements of Geometry* [8] an educational treatise on elementary geometry that takes up some of the contents of the Euclidian *Elements* but is conceived in a post-Cartesian spirit, the Great Arnauld[2] (1612–1694) still believes himself obliged to justify the possibility of multiplying inhomogeneous quantities:[3]

> It is commonly believed that the various kinds of magnitudes, which are called heteroge-
> neous, cannot be multiplied. This does not seem to me to be true, or needs explanation; for
> numbers are of a different kind than other magnitudes such as extent and time, and yet it
> is clear that numbers multiply all kinds of magnitudes and that it is a true multiplication,
> when I say 6 metres or 6 hours, since it is taking one metre or one hour as many times as
> there are units in 6, which is what multiplication consists of.
>   Moreover what cannot be multiplied by nature, can be multiplied by a fiction of the
> mind, by which the truth is discovered as certainly as by real multiplications. Thus, wanting
> to know what distance will make in ten hours someone who has made 80 kilometres in 8
> hours, I multiply by a fiction of the mind 10 hours by 80 kilometres, which gives me an
> imaginary product of hours and kilometres of 800 which being divided by 8 hours gives
> me 100 kilometers. One also multiplies by the same fiction of mind surfaces by surfaces,
> although this gives for product a size of 4 dimensions which cannot be in nature, and
> nevertheless one discovers many truths by these kinds of multiplications.
>
> Arnauld et al. [8, p. 215]

The reading of Arnauld shows it: the best minds have been concerned about these difficulties which, beyond numbers, are inseparable from the astonishing efficiency of mathematics in describing the reality of phenomena. In a few centuries' time, the twentieth-century texts on the problems of interpretation of quantum mechanics will undoubtedly be read with the same somewhat amused curiosity that we experience today when reading of Arnauld's worries about the multiplication of hours and kilometres.

From this point of view, it is emblematic of contemporary blindness to epistemo-logical problems that, in the 1950s, the collective of mathematicians N. Bourbaki was able to justify the absence of an organic explanation of the relationship between mathematics and reality by speaking, in relation to mathematics and physics, of the "purely fortuitous contact between two disciplines" [18]! There is clearly a provocative part to this, a part of ignorance of philosophy and the natural sciences[4] widely shared by several generations of mathematicians in the second half of the twentieth century. More profoundly, these views also echo the concerns of an era fascinated by abstract structures, far beyond mathematics. Since then,

---

[2]Antoine Arnauld, known as the Great Arnauld (*le Grand Arnauld*), is known as one of the main theorists (as a theologian and a logician) of Port-Royal, the school famous for its Jansenism and for Pascal's adherence to its theses.

[3]For convenience, the units of measure in Arnauld's text have been replaced by modern units, rescaled accordingly.

[4]Unlike the great mathematicians of the first part of the twentieth century (Poincaré, Hilbert, Weyl, E. Cartan...), the members of Bourbaki, with a few exceptions, knew little about physics and were openly disinterested in it.

the attitude of the mathematical community has changed profoundly. Interactions with other disciplines are once again seen as a driving and determining factor in mathematical progress. Yet, despite all the interest aroused by these interactions and the development of a literature popularizing mathematical concepts applicable to the study of reality, the epistemological commentaries that accompany them are usually limited to observing the astonishing effectiveness of mathematics when applied to the physical and biological world, and little progress has been made towards a satisfactory epistemological solution to the problem of participation.

The fault may well lie in a methodological error, since the examination of the multiple existing applications of mathematics cannot suffice to settle the de jure question of their conditions of possibility. To this debate, the study of the relationship between "ideal numbers" and "arithmetic numbers", if not exclusive, has however undoubtedly as much, if not more, to contribute as the study of more recent but more complex phenomena such as the mathematization of chaos or quantum theories—two textbook cases from the point of view of the mathematization of reality. It is in any case in this spirit that we will investigate the latter in the following pages.

4.1 The Emergence of the Problem of Participation

The problem and aporias of participation are stated by Aristotle in Book N of *Metaphysics*, where he intends to systematically refute Pythagorism and Platonic idealism by proving that mathematical things do not exist separate from sensible things, and that they are not the principle of things (Aristotle [4, N 6 1093]).

> The first representatives of the doctrine of Ideas, who established two kinds of numbers, the Ideal number and the mathematical number, in no way said, nor could they say, how the mathematical number exists and where it comes from. They made it, in fact, an intermediary between the Ideal Number and the sensible number. [...] All these theories are irrational, they fight each other, and are contrary to common sense.[5]
>
> Aristotle [6, N 3 1090]

Besides, how, if the numbers themselves are in no way found in sensible objects, would the determinations of numbers be found there? (Aristotle [4, N 3 1090])

Plato himself, while developing the theory of Ideas, was acutely aware of the theoretical problem posed by participation: the aporetic character of many Platonic dialogues is too often forgotten. The old Parmenides shows the young Socrates how difficult this theory of Ideas is, even if the idea that there are "forms of Beings" is the only one capable of founding science and knowledge. The reason is always the same. If we admit that Ideas are autonomous with respect to reality, as the roots of

---

[5]Annas [4] translates: "The first people who set up two sorts of numbers, Form numbers and mathematical number, said nothing at all, and cannot say, as to how and where mathematical number is to exist. They put it between Form number and perceptible number [...]."

our knowledge are always empirical, we only know them through their incarnations. Direct access to Ideas and their knowledge is therefore forbidden to us (Plato [95, 134–135]). The theory of mathematical knowledge thus seems doomed to choose between a radical empiricism (where all knowledge is inductive and must be led back to experience) and a pragmatic attitude. The latter amounts to recognizing a certain form of autonomy for mathematical idealities and leaving aside the question of the cognitive roots of the relationships between reality and Ideas and, in the contemporary era, between reality and axiomatic or formal systems. The question has deep didactic ramifications: a large part of the ability to do mathematics lies precisely in the faculty of recognizing in a statement (the problems of trains crossing each other and of bathtubs emptying from elementary school) the occurrence of a theory or abstract techniques (linear algebra, in this case).

Frege's work, with its programmatic idea of a system of pure laws of thought and the development of set theory, but also other more recent theories, allow us to make concrete progress on the phenomenon of participation by leading it back to fundamental cognitive mechanisms. These ideas will be discussed later; for the time being, we will limit ourselves to pointing out three types of response attempts that have become classic, in increasing order of epistemological depth (and not in order of chronological appearance).[6]

4.2   The Logico-Semantic Approach

The first tentative answer, elaborated under various influences,[7] is logical. Recent, it is essentially descriptive, in accordance with the classical thesis according to which logic is all about the form of thought, never about its contents. It codifies the relations of mathematical theories to their empirical realizations or to their applications to the natural sciences, but says nothing about the origins of these relations. It is the theory of models or mathematical semantics: a mathematical theory is seen as a formal system devoid of meaning. A model of this theory consists of an interpretation of the terms, relations and statements of the theory.

Euclidean geometry, for example, admits an axiomatic presentation. Euclid's original presentation was not pure: at times it resorted to the intuitive meaning of certain geometrical operations. Hilbert was the first to give, at the end of the nineteenth century, a rigorous and autonomous formalization of it, without recourse to intuitions or implicit properties, drawing heavily on previous works, such as those of the Italian school. He insisted, and this made even more of an impression than the technical content of his work, on the fact that his new presentation was entirely independent of any reference to the meaning of the constitutive elements of the

---

[6]Frege's ideas and the recent mathematical developments that will be discussed later naturally relate to the third of these, the Kantian theory of schematism of concepts of understanding.

[7]Philosophical, with the ideas of the Circle of Vienna, more technical, with Tarski (1902–1983)....

system. Euclidean geometry as we practice it, by drawing figures to demonstrate Pythagoras' theorem rather than by manipulating formal statements according to formal rules of deduction, is thus an interpretation or model of the Hilbertian system of axioms, of which other models are possible (Hilbert [60]).

This idea applies as such to numbers: we can axiomatize arithmetic[8] in various ways, more or less directly, and conceive our empirical use of numbers as an interpretation of these systems of axioms. Mathematical objects and concepts would thus be purely formal structures devoid of meaning and content; it would be their use in concrete theories (geometric, for example) that would give them meaning and cognitive depth.

The theory of models has an indisputable technical usefulness and relevance. Moreover, historically, it has a philosophical usefulness and legitimacy since it has made it possible to clarify the idea of semantics from the point of view of formal languages. However, its epistemological limits are just as obvious if one takes the trouble to translate its principles into concrete terms, since the notion of interpretation (of a formal system through a model) is basically nothing more than the formalization in the context of axiomatic theories of the notion of participation. Model theory thus provides a language to talk about the relationships between theoretical objects and their possible empirical uses, but says almost nothing about the ultimate reasons of these relationships and the effectiveness of mathematics in the apprehension of the real.

4.3 Nominalism

The second attempt to answer this question has its roots largely in the medieval debates on the mode of existence of ideal objects and, more precisely, universals— those concepts under which fall classes of beings or objects. The central question of nominalism is to decide whether universals (whose fundamental property is to subsume individuals under a common term) have a real and separate existence. The nominalist answer is that it is not the case, the universal being only a simple term: "man" does not exist, there are only human individuals.

Connes' philosophical positions and the idea that "mathematical objects have an existence just as firm as external reality" can be linked to these debates. Nominalism would indeed be resolutely against these views by anchoring the existence of mathematics in language and its conventions. Nominalist theory itself is complex and multifaceted, as are all the philosophies and epistemologies that run through the history of thought. It reached its apogee with the quarrels of medieval metaphysics. In recent times, the positions of the philosophies of language, which seek to lead thought back to linguistic games and conventions, could undoubtedly be interpreted as avatars of nominalist theses.

---

[8] See Chap. 12 of this book.

To stay here within the framework of the philosophy of arithmetic, we will only mention the work of Condillac (1715–1780). It is one of the most interesting in this great lineage of philosophies of language and illustrates well their springs and potentialities. Condillac's language is that of the French Enlightenment: elegant, simple, clear. It speaks for itself:

> All I can think about at first is clearing up anything that embarrasses me. That's why I'm going slowly at first. That's why I stop for a long time on questions that calculators never imagined they should deal with, because these questions are metaphysics, and calculators are not metaphysicians. They don't know that algebra is only a language; that this language has no grammar yet, and that only metaphysics can give it one. [...] To consider numbers in a general way, or as applicable to all the objects in the universe, is the same as not applying them to any of these objects in particular: it is the same as abstracting or separating them from these objects, to consider them separately; and then we say that the general ideas of numbers are abstract ideas.
>
> But, when the ideas of numbers, first glimpsed in the objects, and then in all the objects to which we apply them, become general and abstract, we no longer see them either in our fingers or in the objects to which we cease to apply them. Where then do we see them?
>
> In the names that have become the signs of numbers. Only these names remain in the mind, and it is in vain that one would look for something else [...]. If you believe that abstract ideas are something other than names, say, if you can, what is this other thing? Indeed, when you have disregarded the fingers and other objects which may represent numbers, when you have disregarded the names which are other signs of them, in vain you will look for what remains in your mind, you will find nothing, absolutely nothing.
>
> Condillac [30]

Condillac's nominalism, like modern philosophies of language, has the undeniable merit of proposing a strong philosophical thesis—if nothing exists apart from names, language conventions, if everything is ultimately played out in reality and not in a universe of concepts and ideas, mathematics is only a language, and the problem of participation a false philosophical problem.

4.4 Schematism of Concepts

The last attempt, the most profound but also the most difficult, directly confronts the problem of participation and the relationship between reality and ideals, between objects and concepts, intuitions and abstract theories. It is in Kant (1724–1804) that this theory is built most systematically, under the name of schematism of the concepts of understanding. Mathematics and numbers play a key role in this theory, by illustrating the concrete meaning of schematism.[9] Paradoxically, it is to Heidegger (1889–1976), a philosopher with a reputation for being difficult,

---

[9]The Kantian conception of mathematics, built around the idea of a spontaneous adaptation of arithmetic and geometry to the pure forms of intuition, is also flawed. The mathematical developments of the nineteenth century have, for the most part, invalidated it. We retain from Kantism and the theory of schematism only those contents that remain relevant.

that we owe the most masterful and clear exposition of the theory of schematism (Heidegger [58]). We will largely follow his interpretation.

The fundamental problem, Parmenidean, is that of the application of pure concepts to experience. Now, "in any subsumption of an object under a concept, the representation of the first must be homogeneous to that of the second, that is to say that the concept must contain what is represented in the object to be subsumed". Since empirical concepts are drawn from experience, they are homogeneous to the objects they determine and their application poses no problem. However, "pure concepts of understanding compared to empirical intuitions, or, generally, sensory ones, are quite heterogeneous to them, and could never be found in any intuition. How then is the subsumption of these intuitions under these concepts possible?"

The Kantian argument essentially reproduces the structure of Parmenides' objections. The radical inhomogeneity of the Idea (the ideal number) to the reality (the empirical number) makes the mechanism of participation/subsumption extremely problematic. It is the whole point of schematism to solve the problem, Kant drawing attention to the duality implicit in the phenomenon of subsumption: there is a difference between putting objects under concepts (phase of subsumption proper) and putting into concepts (genetic, synthetic, construction phase).

These two distinct moments are inseparable in mathematical practice and undoubtedly relate to two dual aspects of intuition: recognizing a conceptual schema at work in a phenomenon in the first case (recognizing that the calculation of the moment when two trains cross each other leads back to a standard computation in linear algebra); guessing, in the second, the conceptual schema, the "mathematical structure" underlying a class of phenomena. Heidegger concludes his work on Kant and schematism, one of the most beautiful philosophical texts of the twentieth century, by showing that imagination[10] is what remains essentially unthought of in Kantian theory.

Together with imagination, whose role is well emphasized by Heidegger, the other key element of schematism is the idea of rule, of process:

> It is the representation of a general process of the imagination, serving to give a concept its image, which I call the schema of this concept [...].
>
> The schema must be clearly distinguished from the image. Thus, when I place five points in succession, this is an image of the number five. On the contrary, when I only think of a number in general, which may be five or one hundred, this thought is rather the representation of a method used to represent a quantity (for example a thousand) in an image, in accordance with a certain concept, than it is this image itself. In the latter case, it would be difficult for me to go through the image with my eyes and compare it with my concept.
>
> Kant [79]

---

[10]Imagination is to be understood here in a broad sense, ranging from the production or reminiscence of images (imagination understood in a naive sense) to transcendental imagination in the Kantian sense, which accounts, for example, for phenomena of subsuming intuitions under concepts (as when we draw a triangle or any other figure on a board to try to solve a geometric problem). The quotation that follows, in the body of the text, partly explains this use of the term "imagination" and its relationship to schematism.

In other words, a concept (pure, non-empirical) is inseparable from the rule (the schema) that governs its use: at the limit, it is none other than this rule or, more precisely, by pushing further to the limit the logic of schematism, it is none other than the principle of this rule.

This Kantian interlude allows us to deepen some of the analyses made at the beginning of this book. This functioning of pure, abstract, theoretical concepts, as we find them in mathematical practice, clearly shows, with all due respect to Lichnerowicz, the structuralist school or to logicians and formalists of all kinds, that it is wrong to believe that mathematics is resolutely "non-ontological". In every mathematical concept is genetically inscribed the possibility of its use, and certain intuitive mechanisms that govern this possibility. We will come back to this in the last chapters of this book, showing how certain mathematical ideas such as "universal problems" make the construction of certain mathematical objects (genetic phase) inseparable from their properties and the rules of their use (phase of subsumption, of participation)—a way of using the tools of modern mathematics to try to give a more tangible and effective content to the theory of schematism.

4.5   Schematism and Pragmatism

In the light of the Kantian analysis, it would be natural to identify a concept with its rules of use and imagine that this would make it possible to dispense with technical developments and fairly sophisticated mathematical ideas, such as those just referred to, to solve the problem of participation. The idea is not enough, because pragmatism ignores the intrinsically theoretical nature of mathematics, but it is far from being devoid of legitimacy in many situations, and mathematical history abounds with concepts whose rules of use were well known before the corresponding concept was defined per se. Thus, all modern algebra was constituted, at the beginning of the twentieth century, in a movement of abstraction, associating abstract entities to objects of common mathematical use: to groups operating on a set which had appeared in geometry or in the resolution of polynomial equations (Lagrange (1736–1813); Galois (1811–1832)...) one associates abstract groups; to number fields,[11] the abstract notion of field, etc.

This movement of abstraction, idealization and conceptualization is inseparable from modern mathematics, but its meaning, and the impossibility that goes with it of limiting the scope of schematism to an exclusive recourse to the idea of rule, are well illustrated by the elementary example of unity and its relation to numbers.

The aporias linked to the One and to unity are the subject of Book I of the *Metaphysics*. Aristotle develops there ideas already exposed previously, but also specifies that "one must be careful not to confuse the question of knowing what

---

[11] Sets of real or complex numbers defined by algebraic relations and closed under the operations of addition, subtraction, multiplication, inversion.

kinds of things are said one with the question of knowing what is the essence of the One, what is its notion". The possibility of identifying an intimate kinship between a concept and its rules of use cannot indeed be made in a naive way: this would be, in the language of the classical theory of knowledge, to confuse an extensional definition and an intensional definition.[12] In the case of the One, we already know that things can be said to be one in multiple senses: unity of the concept and functional unity; unity of a totality; topological unity; unity of individuals and universals...Each of these meanings of the word unity can appear in a given mathematical use.

The rules for the use of the concept of unity thus have the diversity of its possible meanings and cannot dispense with the search for the essence of the One, even if we conceive of it as a "rule of rules". The indications given by Aristotle on the nature of the One are not conclusive, but allow us to grasp one of its properties: universality. If there is one remarkable feature of unity—shared in part by number and the idea of number—it is its universality, since "finally, to be one is to possess individual existence". The idea of unity is thus inseparable from the very existence of the world (physical and intellectual) and from the phenomenon of individuation, whether at the level of perception or thought. The different modalities of manifestation of unity (functional, conceptual...) are, in the end, only variations around this fundamental and ontological characterization of "being one", a characterization that goes far beyond the question of "what kinds of things are said to be one".

Let us insist on it: however far they may seem from the descriptions frequently given, these ideas and phenomena are constitutive of mathematical thought. They alone can justify its real functioning. The universality of the concept and the rather mysterious magic of the imagination as it is revealed in schematism are manifested each time mathematical thought has to invent new forms, new methods. Topological unity, for example, is at the origin of calculation methods such as that of the "continuous components" (nowadays we speak rather of connected components) of a space, and is also at the origin of more complex calculations such as those of homotopy groups,[13] which make it possible to classify spaces using algebraic invariants. A mathematics student, when encountering these notions for the first time, will associate them with various numerical intuitions, and will perhaps guess

---

[12] An extensional definition is obtained by listing all the objects that fall under a concept (we can thus define precious stones by listing the stones considered as such: rubies, emeralds...). An intensional definition is based on conceptual determinations (with, for example, the definition of man as an animal provided with reason). However, modern mathematics has shifted and reconfigured the classical dividing line between the two notions by giving the theoretical legitimacy of conceptual determinations to certain extensional definitions—at the price of a methodological systematization of the latter, through notions such as equivalence relations.

[13] It does not matter here how they are defined. The first homotopy group, also called Poincaré's group, appears for example when one tries to count (by an abstract approach) the number of holes in a planar surface or (in a roughly equivalent way) to study the monodromy of differential systems (that is, for example, the behaviour of the solutions of differential equations on the complex plane when the coefficients of the equation, functions of a complex variable, contain singularities—tend towards infinity for certain values of the variable).

that one can calculate with spaces as one calculates with numbers: one can add spaces (by disjoint union), multiply them (by "Cartesian products"), and even define an exponential function. Only the extraordinary flexibility of the rules of imagination (Kantian, conceptual), inseparable from the problem of participation, makes it possible to understand this creative ability of thought and its faculty to extend from the domain of numbers to that of spaces the methods of algebraic calculation.

Chapter 5
The Third Man Argument

The third man argument is one of the keys to the classical and modern understanding of numbers. We owe its introduction to classical Greek philosophy. The argument reappeared in the nineteenth century, but the mathematicians of the time do not seem to have realized that they were putting arguments and debates from antiquity back on the agenda. Its implications in modern mathematics are manifold and are mainly based on the works of Bolzano (1781–1848), Dedekind (1831–1916), Cantor (1845–1918) and Frege with, in particular, the existence and nature of infinity and a "logical" construction of integers.

Philosophical tradition agrees that the origin of the third man argument is to be found in Plato's *Parmenides*. The old Parmenides, we remember, tries to show the young Socrates the theoretical difficulties intrinsic to the method of Ideas. One of the central problems, already encountered, is the inhomogeneity between objects and Ideas, one of the recurring temptations of Platonic idealism being to make ideal objects (the ideal number...) objects 'like the others'. The argument of the third man (which takes the form of an argument of the 'third greatness' in the *Parmenides*) aims to show its intrinsic aporias:

> (Parmenides): Here, I think, is what makes you judge that each idea is one. When several objects seem great to you, if you look at them all at once, it seems to you that there is in all of them one and the same character, from which you infer that greatness is one.
>
> – True, says Socrates.
> – But if you embrace in your thought both greatness itself and great things, do you not see yet another greatness appearing, by which all these necessarily seem great?
> – It seems so.

© The Editor(s) (if applicable) and The Author(s), under exclusive license
to Springer Nature Switzerland AG 2020
F. Patras, *The Essence of Numbers*, Lecture Notes in Mathematics 2278,
https://doi.org/10.1007/978-3-030-56700-2_5

– So then another form of greatness will appear, alongside greatness itself and the things
that participate to it, and then, over and above all those, still another, by which they are
all great. Thus, each of your forms will no longer be one, but infinite in number.[1]

Plato [94, 131–132]

The Idea, the form, gives an account of what certain things have in common. If
it behaves itself as one thing, it obviously has some features in common with those
things which it accounts for. One can thus form the Idea of man (let us call it the
man-in-itself Idea), then the Idea of the "third man", which gives an account of what
is common to individual men and to the man-in-itself. Then one could similarly form
the Idea of the "fourth man", which would account for what is common to individual
men, to the man-in-itself and to the third man, and so on, to infinity—an idea which
Greek thought is repugnant to.

Aristotle uses the argument several times to show that of all the dialectical
arguments by means of which the Platonists claim to demonstrate the existence of
Ideas, none is obvious.[2] Indeed, the paradox seems to be rooted in the nature and
functioning of thought, which is reluctant to create useless concepts: to the question
"what is human?" we will answer both man (the man-in-itself, abstract, general)
and this man, short-circuiting the iterative mechanism suggested by the model of
the *Parmenides*.

5.1   Any and Everything

The argument of the third man admits a second classical interpretation, due to
Alexander of Aphrodisius [1, (150–215, approx.)]:[3] if I say "the man walks" (in
the sense of a general statement like "the deer bawls", "the dog barks"), the man
in question is neither this particular man (Socrates), nor the Idea of man (unable
to walk!). There is thus, in between the objects and their Idea, a third term: here, a
new "third man", which would be something like "the generic man" rather than "the
man-in-itself".

---

[1]Allen [95] translates: "I suppose you think that each character is one for some such reason as
this: when some plurality of things seems to you to be large, there perhaps seems to be some
one characteristic that is the same when you look over them all, whence you believe that the
large is one.—True, he said.—What about the large itself and the other larges? If with your mind
you should look over them all in like manner will not some one large again appear, by which
they all appear to be large?—It seems so.—So another character of largeness will have made its
appearance, alongside largeness itself and the things which have a share of it; and over and above
all those, again, a different one, by which they will all be large. And then each of the characters
will no longer be one for you, but unlimited in multitude."

[2]Aristotle, *Metaphysics* [4, M 4 1079a] and [6, A 9 990b].

[3]This interpretation could already be underlying Aristotle, *On Sophistical Refutations*, [2, 178–
179].

From the point of view of arithmetic, a statement like "the Number is even or odd" makes little sense: the Number is not a particular number. If the statement had a meaning, one should be able to apply to it the rule of the excluded third: the Number would be either even or odd—a type of aporia which runs through all the literature on numbers and mathematics, largely due to the ambiguities of non-formalized languages. Mathematics accommodates this difficulty by introducing in its reasoning the idea of generic number, which allows to instantiate the idea of number while keeping its generality: "Let n be any number, then n is even or odd." Since Frege [52] and Peano (1858–1932) [93], these two orders of statements are distinguished through the notion of free or bound variable. The correspondence with the distinction at work in the third man argument in Alexander of Aphrodisius' version is only imperfect, but the post-Fregean point of view has the undeniable merit of clarifying the problem.[4]

Bertrand Russell (1872–1970) provided its clearest conceptual and methodological exposition. He traced back the distinction to Euclid. It deserves attention because it illustrates an essential mechanism in mathematical thinking.

> Given a statement containing a variable x, say "$x = x$", we may affirm that this holds in all instances, or we may affirm any one of the instances without deciding as to which instance we are affirming. The distinction is roughly the same as that between the general and particular enunciation in Euclid.[5] The general enunciation tells us something about (say) all triangles, while the particular enunciation takes one triangle, and asserts the same thing of this one triangle. But the triangle taken is any triangle, not some one special triangle; and thus although, throughout the proof, only one triangle is dealt with, yet the proof retains its generality. If we say: "Let ABC be a triangle, then the sides AB, AC are together greater than the side BC", we are saying something about one triangle, not about all triangles; but the one triangle concerned is absolutely ambiguous, and our statement consequently is also absolutely ambiguous. We do not affirm any one definite proposition, but an undetermined one of all the propositions resulting from supposing ABC to be this or that triangle. This notion of ambiguous assertion is very important, and it is vital not to confound an ambiguous assertion with the definite assertion that the same thing holds in all cases.
>
> Russell [100]

Russell, following Peano, calls the variables of a propositional function real or apparent depending on whether they correspond to the case "any" or "all":

> [The distinction] was, I believe, first emphasized by Frege. His reason for introducing the distinction explicitly was the same which had caused it to be present in the practice of mathematicians; namely, that deduction can only be effected with real variables, not with apparent variables.
>
> Russell [100]

---

[4]Other approaches would be possible, such as the use of Hilbert's symbol, to address this idea of genericity. See e.g. Bourbaki [17, I 16]. On Hilbert's mathematics and logic, see Boniface [15] and Cassou-Noguès [29].

[5]For example, in Book 6, Prop. 1 of the *Elements*: "Triangles and parallelograms which are under the same height are to one another as their bases" and Book 7, Prop. 1: "Two unequal numbers being set out, and the less being continually subtracted in turn from the greater, if the number which is left never measures the one before it until a unit is left, the original numbers will be prime to one another." (Euclid [46]).

The Russellian distinction underlies in fact the whole enterprise of type theory, which will soon be discussed. This is easily understood by going back to the conceptual origins of the problem, but has not failed to puzzle the logical commentators who have minimized its scope, no doubt for lack of understanding of its concrete mathematical meaning. In the reference volume by van Heijenoort [59] on the sources of modern mathematical logic, which includes Russell's article on type theory, Willard Quine (1908–2000) proposes a critical interpretation of the article that is indisputable in its technical aspects but leaves one sceptical as to its conceptual relevance. By dissociating mathematical logic from its operative and even ontological correlates, philosophy can easily lose its relevance and credibility among mathematicians. According to Quine,

> Russell is thus led to propound a distinction between "all" and "any": "all" is expressed by the bound ("apparent") variable of universal quantification, which ranges over a type, and "any" is expressed by the free ("real") variable, which refers schematically to any unspecified thing, irrespective of type [...]. This contrast between asserting a universal quantification and asserting a "propositional function" carries over into the first edition of *Principia Mathematica* and puzzles readers who do not perceive that it is pointless apart from certain aspects of the theory of types.
>
> van Heijenoort [59, p. 151]

Quine's assertion, influenced by technical problems and the subsequent evolution of mathematical logic, is all the more surprising since Russell explicitly insists on the mathematical function of his distinction, and on its inescapable role in mathematical work. There is always, for a mathematician, a feeling of uneasiness when reading authors such as Quine, Carnap (1891–1970) or Wittgenstein:[6] they very regularly talk about mathematics and claim to account for it, but with a discourse and an approach that are frequently out of step with the reality of mathematical work, its methodology and its modus operandi.

What is most interesting for the philosophy of mathematics and number, in the idea of a real or generic variable, is its ontological status: as with Alexander of Aphrodisius, as in the assertion "man walks", it has both the generality of the concept and the properties of concrete (mathematical) objects. The particular statement in Euclid and its proof (for example about a generic triangle ABC) is fascinating in that the proof relating to this triangle (which can for example be drawn on the board, an operation in which it loses its generality) is susceptible of as many instantiations as there are triangles (an infinity). It is this possibility of instantiation, largely implicit when a teacher gives the proof of Pythagoras' theorem to young children on the blackboard, that ensures the universal validity of the theorem. It is tacitly based on the idea (again rarely made explicit) that the proof depends in fact only on the nature of the object under study (the triangle in this case), so that even if the teacher ultimately chooses a specific instantiation on the board, the proof

---

[6]Representatives of analytic philosophy. This feeling of unease does not exist when reading their precursors, such as Frege, and only begins with Russell, in whom, despite all the talent and great scientific intuitions, mathematical sensitivity is already beginning to weaken under the influence of logic.

loses nothing of its generality. It is clear here that, although general and particular statements in the Euclidean sense are apparently equivalent, there is a profound difference in terms of knowledge theory—they refer to two ways of constructing theoretical content.

The logician and philosopher Giuseppe Longo[7] has studied these phenomena from the point of view of contemporary logic.[8] He illustrates them, among other things, through an example attributed to the young Carl Friedrich Gauss. It consists in calculating the sum of the first n positive integers: $S(n) = 1 + 2 + \ldots + n$. Gauss proceeds by a geometrical argument, by matching the first n positive integers in pairs:

$$S(n) = 1 + 2 + \ldots + (n - 1) + n$$
$$= n + (n - 1) + \ldots + 2 + 1,$$

so that $2S(n) = (1 + n) + (2 + (n - 1)) + (3 + (n - 2)) + \ldots + (n + 1) = n(n + 1)$ and $S(n) = n(n + 1)/2$. The fascinating thing about this argument is that it simultaneously implements several cognitive phenomena that a classical logical and formalist approach is unable to account for. First of all, the argument is underpinned by a geometrical representation of the sequence of numbers. This representation is itself generic and idealized: n being an arbitrary number, the dotted lines are there to indicate that an exact representation of the sequence (which would suppose a number of fixed terms, thus the choice of a precise value for n) is impossible. In a second moment, the argument implements a process of dynamic association whose manifestation is found in many combinatorial techniques and which implements, in addition to the generality of n, notions of order and symmetry.

All of Longo's work suggests that these phenomena can be traced back to the very mechanisms of cognition, "implemented" in the mathematician's brain by evolution, starting with the cognitive proto-experiments of space, movement perception and kinestheses. It also indicates that modern attempts to base all mathematics on arithmetic, with a very strong emphasis on the role of induction,[9] have missed such fundamental phenomena as the role of genericity (and the associated proofs, the "prototype proofs" (Longo [83, 84])). As a consequence, the mathematical epistemology of the twentieth century largely underestimated, in favour of formal

---

[7]From whom I have learned the epistemological importance of the notion of genericity. I thank him here in a friendly manner. G. Longo also emphasizes the fundamental difference between the genericity of a variable in algebra, in its domain of variation, and geometric genericity, which is that of a triangle or categorical diagram, whose genericity must be demonstrated once it has been drawn and used in a proof. The ramifications of these ideas in contemporary science are further explored in the direction of life sciences in Bailly and Longo [9].

[8]Longo [83] and [84].

[9]If an assertion about positive integers n, $P(n)$, is true for $n = 1$ and if the implication "$P(i)$ implies $P(i + 1)$" is true for any positive integer i, then $P(n)$ is true for any positive integer n.

proofs and deductive schemes, the role of the construction of concepts and structures in mathematical progress.

5.2 The Infinity of Ideas

The argument of the third man in its original form, that of the *Parmenides*, was to reappear in a different context throughout the nineteenth century. Let us recall the conclusion of the *Parmenides*: if we accept the theory of Ideas, "each form will no longer be unique, but infinite in number". The aporia thus branches out in two directions: can Ideas be treated as objects; what are the logical and ontological consequences? Let us recall that during the Late Middle Ages these questions were considered central, among other things because of nominalism.

While it is difficult to determine precisely what conceptual influences may have been at work throughout history at the interface of philosophy, mathematics and theology, it is not very surprising in retrospect that the great re-inventor of the third man argument in the early nineteenth century was Professor of Philosophy of Religion at the University of Prague. Bernard Bolzano (1781–1848), now considered one of the founders of set theory and of the arithmetization of analysis, was indistinctly a mathematician, logician, philosopher, and theologian. His *Paradoxes of Infinity* [14] inherited the multiplicity of his interests and are both philosophical and mathematical, as will be the work of Georg Cantor after him. The central question is the mode of existence of the infinite:

> Does this concept [of infinity] have an objective existence, that is, are there things of which it can be the attribute, are there sets that we can rightly call infinite? I will answer resolutely affirmatively. In the realm of things that do not claim to exist actually nor even only to be possible, there are unquestionably infinite sets. As it is easy to see, the set of all propositions and truths in themselves[10] is infinite. If we consider, indeed, any truth, for example the proposition "there are truths in general", or any other proposition that I label by A, we notice that the proposition "A is true" is different from A itself; the latter obviously has a different subject than the former, the subject of the former being the entire proposition A. Let us designate by B this second proposition: "A is true", and let us repeat the derivation process that has already given us B from A, we will obtain a third proposition C from B, and so on and so forth indefinitely [...]. The collection of all these propositions is greater than any finite set. The reader will notice the similarity of this sequence of propositions with the sequence of numbers.
>
> Bolzano [14]

---

[10] A proposition in itself (*Satz an sich*) is, according to Bolzano, a statement that something is or is not (Bolzano [14, n. 31]).

The same argument which constituted an objection for the Greeks, the fact of arriving at an infinity (of ideas or propositions), thus became in the nineteenth century a proof of the existence of infinity in act.[11]

At the end of the nineteenth century, Dedekind no more than Bolzano before referred to Greek origins for his famous demonstration of the existence of infinite sets. If it seems rather improbable that, as a mathematician in his own right, he was directly inspired by philosophical or metaphysical sources,[12] one notes in his work a proximity with the argument of the *Parmenides* that is even more striking than in Bolzano, in whom the recursive construction took a semantic form ("A is true", "'A is true' is true",...) rather than a Platonic one (The Idea, The Idea of the Idea,...). If Dedekind's proof rests on a determination of infinite sets that had already appeared in Bolzano ("A system is said to be infinite when it is in bijection with one of its own parts; in the opposite case it is said to be finite"),[13] its construction of an infinite set rests entirely on the argument of the third man:

Proposal: There is an infinite system.

Evidence: The world of my thoughts, that is, the totality S of all things that can be the object of my thought, is infinite. For, if s is an element of S, so is the thought s', that s can be the object of my thought. If we see s' as the image $\varphi(s)$ of the element s, the transformation φ thus determined of S in itself has this property that its image s' is a part of S; and it is indeed a proper part ($S \neq S'$), because there are elements (for example my own self) in S which are necessarily distinct from all s' and cannot therefore belong to s' [...]. Thus, S is infinite, QED.

Dedekind [36]

Dedekind's argument, which some consider to be the founding father of modern algebra,[14] is surprising in many ways. The reference to the "self" as the first element of the system has an obvious kinship with the Cartesian cogito, but is more likely to reflect philosophical debates at the end of the nineteenth century, where subjectivity played, in multiple forms, a driving and founding role for the whole theory of knowledge (with Brentano (1838–1917), Husserl, or Freud (1856–1939)).

---

[11]In fact, Bolzano—no more than Dedekind after him—does not leave the realm of the potential infinite. Zermelo's work has shown that the existence of an infinite set cannot be demonstrated, see Bolzano [14, n. 33] and Boniface [15, p. 18].

[12]Dedekind is in fact said to have been inspired by Bolzano, see Boniface [15, p. 18, n. 2], although opinions differ on the matter (Gericke [55] seems rather to follow Dedekind's testimony that his work "was carried out at a time when the work of Bolzano and its very name were completely unknown to him").

[13]Bolzano [14, §20].

[14]Corry [32].

5.3   The Recursive Construction of Numbers

The agreement of the Dedekind and Bolzano method with the Parmenidean mechanism is not perfect however: in Plato, Parmenides does not indeed consider the Idea of the Idea... of the Idea of Greatness, but, in a more subtle way, starting from a class of objects O (the great objects), forms their Idea (the greatness, I1), then forms the Idea I2 of the group formed by I1 and the primitive class O, then the Idea I3 of the group formed by O, I1, I2, and so on. It is precisely this scheme which was retained in the twentieth century to serve as a foundation for a recursive definition of numbers whose invention (brilliant, and which goes far beyond the sole use of the Parmenidean recursion) is due to Frege. We will explain it in detail later.

Frege's seminal text [50] for the theory of number, the *Foundations of Arithmetic*, appeared in 1884, 4 years before Dedekind's.[15] It should be noted that Frege avoids a pitfall that Dedekind had not known how or did not want to avoid by basing his construction on the "self". Indeed, Frege chose as a starting point for the construction of integers not a particular object or class of objects, but the concept "distinct from oneself" (Frege [50, §74]). This vocabulary of concepts[16] can be translated into the language of set theory (to a concept is formally associated the set of objects that fall under the concept, here the set of x's distinct from themselves, i.e. the empty set) or that of propositional functions (in this case the propositional function $x \neq x$). Contrary to Dedekind's "self", Frege's starting point is thus of a purely logical nature and makes no reference to the mathematician's subjectivity.

The Fregean attitude towards the foundations of mathematics would quickly become the norm in the twentieth century, leading to ever-increasing difficulties of communication with the philosophy of the great tradition (metaphysics, Kant...), which, faithful to the Cartesian demand (the ultimate reference to the cogito), was never to give up making the thinking subject the touchstone of the theory of knowledge.[17]

5.4   Type Theory

Finally, the third man argument has yet another mathematical correlate: it may well explain the reluctance that mathematicians may have quite naturally towards the type theory[18] of Russell (1872–1970) and Whitehead (1861–1947) which, quite

---

[15]Dedekind was very late in publishing it; it was actually written before the publication of the *Foundations of Arithmetic*.

[16]Frege's choice to reason in terms of concepts is anything but arbitrary: see Chap. 10 of this book.

[17]See e.g. Husserl [67].

[18]Here we consider the initial type theory, which had a strong epistemological ambition, in the Fregean tradition. There are various later variants, useful in logic or computer science. The problems we raise are related to the existence of a hierarchy of types. They can be partly

faithful to the Parmenidean idea of a stack of ideas of ideas of ideas, has proved to multiply mathematical objects unnecessarily. Contrary to Russell's initial opinion [100] that "the theory [of types] also has, if I am not mistaken, a certain agreement with common sense, which makes it intrinsically credible", it does indeed seem that thinking (mathematical and conceptual) is not "typed" in Russell's sense, which allows us to treat distinct conceptual levels (the man-in-self, this man...)[19] in a uniform way. We will only briefly report here on these aporias of type theory,[20] insisting only on their analogy with those concerning the third man and referring for the moment in a rather allusive way to phenomena belonging to set theory.[21]

Russell's objective, when he introduced type theory in 1908, which was taken up again in the *Principia Mathematica* with A.N. Whitehead [111], was to solve the too many paradoxes that mathematical logic and set theory were then facing. The contradictory character of the set of sets (Cantor, 1899), the impossibility of a set formed of all ordinals (Burali-Forti, 1897), or Richard's paradox[22] are some known examples—Russell gives a commented list of paradoxes, accompanying it with solutions written in terms of type theory ([100]).

All these contradictions, he remarks, have a common characteristic, well illustrated by the oldest of the paradoxes of this family: the paradox of the liar. According to Epimenides the Cretan, all Cretans are liars, and everything they say is false. In another form, the man who says "I am lying" can neither lie nor tell the truth. The essence of Epimenides' sentence and other contradictions is that "something is said about every case of a certain kind. But the paradox itself seems each time to give rise to a case which, at the same time, is and is not one of the cases considered": self-referentiality, the vicious circle, is the only source of contradiction. A few years after Russell and Whitehead, Weyl (1885–1955), who, moreover, does not share their conclusions, will thus analyze the situation of analysis at the beginning of the twentieth century:

---

circumvented by ad hoc axioms which however reduce the philosophical scope of the typed theories.

[19] It is important to be precise here, as this is a delicate point. Russell is right in that mathematical thinking is spontaneously typed, in the sense that we spontaneously classify mathematical objects: a triangle is a geometric object of a particular type (a figure, a polygon...). However, this typification is fluctuating, and this is one of the riches of mathematical thought, analyzable in terms of horizon structures in the context of Husserlian phenomenology. As we will see, Russell's theory, despite its intuitive background, rigidifies this spontaneous typification too much for his type theory to correctly describe mathematical practice.

[20] For a more technical analysis, see Potter [99, chap. 5: Russell's account of classes].

[21] We consider here the original type theory, conceived in the spirit and logic of set theory and its aporias. Modern type theories, as used for example in theoretical computer science, have integers built in and are compatible with the practice of mathematicians. On the other hand, they no longer claim to construct integers as cardinals, as was the case with Russell.

[22] The set of integers whose definition is expressed in less than sixteen French words is finite, and yet it is contradictory to define an integer as "*le plus petit entier qui n'est pas définissable en moins de seize mots français*" (the smallest integer that is not definable in less than sixteen French words), because the definition has only fifteen words!

The *circulus vitiosus* [. . .] is surely not an easily dispatched formal defect in the construction of analysis. But, the more distinctly the logical fabric of analysis is brought to givenness and the more deeply and completely the glance of consciousness penetrates it, the clearer it becomes that, given the current approach to foundational matters, every cell (so to speak) of this mighty organism is permeated by the poison of contradiction and that a thorough revision is necessary to remedy the situation.

Weyl [109]

The guiding idea of type theory is to associate to any propositional function ("if x is a man, x is mortal") a limitation of the range of variation of its argument (x), outside of which the propositional function has no more values (and thus loses all meaning). The proposition "if 3 is a man, 3 is mortal", acceptable from the point of view of traditional formal logic (3 is not a man, the proposition is therefore formally true), is not acceptable from the point of view of the semantics of ordinary language,[23] nor from the point of view of type theory (where 3 will be an object of another type than man). The essence of the Russellian approach lies in this establishment of a certain logical homogeneity in the formation of propositions, which is lacking in paradoxes.

In practice, a type is "defined as the range of significance of a propositional function, i.e., as the collection of arguments for which the said function has values". An iterative construction is set up: roughly, given an arbitrary domain of individuals (called type 1), one forms the propositions admitting these individuals as "real or apparent" variables (free or bound): these are the propositions of the first order (second logical type). "We have thus a new totality, that of first-order propositions. We can thus form new propositions in which the first order propositions occur as apparent variables. These second-order propositions form the third logical type" (Russell [100]), and so on.

To make sense, a quantification (whether it concerns propositions, propositional functions, classes. . .) must then always specify a logical type. Thus, in the paradox of the liar, a well-formulated statement in Russell's system would be: "All first-order propositions affirmed by me are false." Since it is a second-order proposition (the quantification is over first-order statements), it is not contradictory in its content (I can systematically state false first-order propositions, and state a true second-order proposition claiming that they are false).

With the construction of integers, type theory shows its limits and its fundamental incompatibility with mathematical practice: "It is to be observed that 0, 1, and the other cardinals are [in this system] ambiguous symbols, and have as many meanings as there are types." Indeed, a cardinal number is, by nature, associated with the counting of objects in a class. Since classes are typed, numbers must themselves have logical types, and the principles of logical homogeneity of type theory therefore prohibit treating cardinal numbers in a uniform way. To give a

---

[23] It makes no sense, since numbers on the one hand and humanity and mortality on the other are part of distinct semantic fields that only poetic language can be brought to superimpose—but the poetic effect lies precisely in the transgression of the rules of ordinary language semantics!

schematic idea of the problems which arise,[24] let us suppose that the individuals of the theory are objects of Euclidean geometry: the number of vertices of a hexagon will be of logical type 3. Indeed, in the Russellian system (which takes up the Fregean theory of number, which we anticipate briefly here), a number is a class of equipotent sets.[25] The number of vertices of a hexagon is thus a class (of type 3) formed of sets (of type 2, like the set of vertices of the hexagon) of objects (of type 1, like the vertices themselves, which we consider here as primitive objects of the considered geometrical theory).[26]

The cardinal numbers (the truly "natural" numbers of the system, those obtained by counting the finite sets of primitive individuals of the system) are thus of logical type 3: we denote this by $(n, 3)$ (where n is a number and 3 designates its logical type). It would be expected that x, the "number of non-zero integers strictly less than $(8, 3)$" is equal to $(7, 3)$ (which would translate in a typed language that 7 is the number of non-zero integers strictly less than 8). However, this is not the case: for the same reasons as before, x is a class of object classes of type 3, and is therefore of logical type 5! In short, $x = (7, 5)$. All arithmetic (and all mathematics) becomes impossible, except by introducing ad hoc processes (one could for example decide to work only with infinite types). Weyl's conclusion is indisputable: A "hierarchical" version of the analysis is artificial and useless. It loses sight of its object, i.e., numbers.

All this unnatural imbroglio has led Russell to introduce axioms of reduction to short-circuit typification (so as to neutralize these problems), but these new axioms ruin the logical edifice of the theory, its conceptual economy, and ultimately its raison d'être.[27] This is one of the reasons why, in mathematics, type theory has been supplanted by more economical theories that are more faithful to mathematical

---

[24]For a more precise analysis, we refer to Russell's original article, which is very precise about the ambiguities that arise when typed logic is applied to elementary arithmetic.

[25]I.e. sets that can be matched bijectively: whose elements can be coupled pairwise.

[26]In fact, type theory is a logical puzzle since one could also construct the vertices of the hexagon as elements of an arbitrarily high type (depending on the construction method chosen and on the quantization used in the construction)—suppose for example that one considers points as primitive objects of the system (logical type 1) and straight lines as sets of points (thus corresponding to propositions admitting objects of logical type 1 as free or bound variables: a straight line is then of logical type 2). The intersection of two non-parallel lines in Euclidean typed geometry will then certainly be a point, but of a logical type strictly superior to 1. Logically, one must therefore distinguish a point in space seen as a primitive object and this same point seen as the intersection of two lines.

[27]Weyl, in *The Continuum* [109], will propose another approach, limiting the use of quantifiers to the first logical types. His solution is, to a certain extent, intermediate between Russell and Zermelo–Fraenkel, keeping, from typification, the idea of an iterative construction of mathematics but without resorting to Russell's axioms of reduction. It forces Weyl, in order to guarantee its logical consistency at all costs, to restrict the methods used in the analysis.

practice, such as the Zermelo–Fraenkel system which, by restricting the universe of set theory, bypasses the artifice of logical types.[28]

Note finally the profound analogy between typification and the aporia of the third man. Russell, like Parmenides, comes to distinguish, for the same notion (man, a natural number, any other mathematical object in the theory of types), levels of typification (the Idea of the Idea of..., the cardinal of the cardinal..., men, man-in-itself, the third man...). In both cases, the method is counter-intuitive. In both cases, it produces unnatural objects and is in conflict with the functioning of thought (the Idea of the Ideas of Man is largely an artifice of thought, as are typed cardinal numbers). Here we come up against an irreducible phenomenon, a key to the empirical functioning of thought, no doubt of a complexity comparable to that underlying the schematism of concepts.

Acknowledgments My friendly thanks to André Hirschowitz and Giuseppe Longo for their rereading of this chapter, and their sound advice

---

[28] See for example Bourbaki [17]. In practice, the mathematical community has been little involved in these choices, which largely reflect positions, issues and convictions internal to the field of mathematical logic.

Chapter 6
Numbers and Magnitudes

One of the great discoveries of mathematical thought, whose development extends from Aristotle's work on logic in the fourth century BC to Hilbert's work at the end of the nineteenth century, is the idea of axiomatics. An axiom, or a system of axioms, traditionally (in Aristotle, for example) formulates a general statement, the evidence for which is indisputable. The idea of generality is very important: it is this idea that Hilbert, and then all the algebra of the early twentieth century, put forward when it was noticed that the same system of axioms (like those of group theory) could describe very disparate mathematical ideas (in the case of groups, displacements in space as well as substitutions between the roots of a polynomial equation).

This versatility of mathematical objects can already be found in numbers. It has been at the origin of countless debates, as to the true nature of numbers, as to their cognitive origins and as to their meaning. The key problem is the relationship between numbers and quantities, or numbers and magnitudes. To put it simply: is it the "same" number 5 that we use to say that there are 5 of us sharing an office and that the next subway leaves in 5 min? Of course, such distinctions are obscured in our daily lives by years of indiscriminate use of numbers in different contexts, but the difficulties of young children in acquiring the protean mastery of arithmetic that adults have suggest that this is more than a scholastic difficulty.

As far as our cognitive faculties are concerned, it is quite obvious that the act of counting, of a rather combinatorial nature, is not homogeneous to that of measuring durations or distances, of a rather geometric nature. Nevertheless, it is the same numbers that are used in arithmetic and algebra or geometry and mechanics. To express ourselves in Wittgenstein's way, it is the same rules (of multiplication, addition…) that we use each time we have to use numbers, in contexts that are fundamentally different and that go well beyond the phenomena of counting and measurement. This chapter will be devoted to understanding the modalities of this unit, and the philosophical difficulties it raises.

© The Editor(s) (if applicable) and The Author(s), under exclusive license   57
to Springer Nature Switzerland AG 2020
F. Patras, *The Essence of Numbers*, Lecture Notes in Mathematics 2278,
https://doi.org/10.1007/978-3-030-56700-2_6

6.1   The Notion of Quantity

An even more general question underlies those that can be asked about numbers: what is the purpose of mathematics? This is a difficult question, which embarrasses any mathematician who wants to give an honest answer. "The science of numbers and space" is a classic answer, faithful to the origins of mathematical thinking (arithmetic and geometry), but unsatisfactory in that it says nothing about the concepts common to both disciplines.

Aristotelian logic allows us to go a little further. According to it, ten major categories[1] form the most general genera, i.e. the most universal concepts to which all types of beings, objects and phenomena must be subordinated as soon as they are the subject of scientific thought: substance, quantity, quality, relation, location, time, position, possession, action, passion. For us, moderns who know that physics is mathematicizable, there is probably none of these categories that cannot be given a mathematical content. In Aristotle, two of them can be given a natural such content: quantity (e.g. "long by three feet") and, less clearly stated, relation ("double, half, bigger").

The analysis of quantity in the Categories seems to solve once and for all the problem of relationships between numbers and quantities, Aristotle establishing a number of simple and fundamental conceptual distinctions:

> Quantity is either discrete or continuous. Moreover, quantity consists either of parts having a position with respect to each other or of parts not having a position with respect to each other. Instances of discrete quantities are number and speech; of continuous quantities: the line, the surface and the solid, and, in addition, time and place. With respect to the parts of a number, there is no common limit where the parts are in contact. Thus, five being a part of ten, the two fives do not touch each other at any common boundary; on the contrary, these two fives are separated. [...] Number is therefore a discrete quantity. As for the line, it is a continuous quantity, because it is possible to conceive of a common limit where its parts touch: it is the point, and for the surface, it is the line, because the parts of the surface also touch at a common limit [...]. Time and place are also of this kind. [...] In addition, there are quantities which consist of parts which have a position in relation to one another and other quantities which are not composed of parts which have a position. Thus the parts of the line have a position in relation to one another: each of them is located somewhere [...]. As far as the number is concerned, on the contrary, it would not be possible to show that its parts have a reciprocal position.
>
> Aristotle [5, 4b]

6.2   Number as Measure

The text we have just quoted has had a considerable influence on the philosophy of number and on mathematics itself: the distinction between the discrete and the continuous would thus form a natural and imprescriptible dividing line between

---

[1] Aristotle, *Categories* [*1b*]. The translation of the *Categories* is based on Tricot [5] and Cook [3].

numbers and quantities! Nevertheless, this distinction does not exhaust the subject and Aristotle returns to it in the *Metaphysics* by articulating the idea of unity to that of measurement. The Aristotelian One, a principle of totalization, is what makes things be one, but it is also a regulating principle without which measuring, that is to say, relating to a standard, a measure, would be impossible:

> To be one is to be a kind of origin of number; for a first measure is an origin, for what first makes each genus intelligible to us is its first measure. The origin, therefore, of our acquaintance with each [kind of] thing is that which is one. But that which is one is not the same thing in every genus; for it may be here a quarter-tone, there a vowel or mute, and another thing in the case of weight and something else in the case of change.
>
> Aristotle [7, $\Delta 6$ 1016b]

> For the measure is that by which the quantity is known; but it is by the One or by a number that the quantity as a quantity is known; and every number is known by the One. Therefore every quantity as a quantity is known by the One, and that by which quantities are primitively known is the One itself; and therefore the One is the principle of the number as a number. Hence it is that, in the other categories also, the name of measure is given to that by which primitively every thing is known, and that the measure of the various kinds of beings is a unit, a unit for length, for width, for depth, for weight, for speed.
>
> Aristotle [6, I 1 1052b]

In terms of cognitive faculties, this is the direction in which we have to look for the ultimate reason that convinces us that enumerating and measuring are two facets of the same activity: in both cases, we institute a norm that allows the use of numbers. In the discrete case, the norm is conceptual: counting characters in a Renaissance group picture implies that we have chosen the notion of character rather than another normative notion such as that of notable or ecclesiastic. In the continuous case, the norm is a unit of measurement and is therefore conventional: the selection of a unit of size (the metre, the foot. . .) is the result of a largely arbitrary choice. "This is why the measurement of number is the most exact of all, because the unit is set as an absolute indivisible; all our other measurements are only imitations of it." (Aristotle [6, I 1 1052b])

The example of music, widely debated in Greek thought, is interesting in that it gives an example where a physical and experimental notion (musical pitches, a priori arbitrary and susceptible to continuous variations) inherits from the musical tradition and the physiology of hearing a discrete division into semitones. We know that contemporary music has sometimes tried to free itself from this by moving beyond the twelve-tone system towards other interval theories (quarter-tones. . .). Music theory thus offers a privileged terrain for thinking about the arithmetization of physical phenomena and their anthropomorphic and cultural character: how far, in particular, is the weight of Western musical tradition legitimate? To what extent is challenging it, more or less radically, legitimate?

6.3 One Is Not a Number

The Aristotelian theory of number leads to a surprising statement: one is not a number. There is something scandalous here for us, who are now used to calling "numbers" mathematical beings such as negative, complex or quaternionic numbers. The same scandal is repeated for zero, whose acceptance as a number was long considered problematic. We must therefore try to understand Aristotelian thought and, beyond that, the reasons that, over the course of time, have led several mathematicians, often subtle and experienced, to deny this or that object the status of number.

First of all, the objection that one is not a number is not a practical one: performing an operation such as $5 + 1 = 6$ was obviously legitimate in Greek times. The problem is therefore not technical, contrary to some of the objections, which will come much later, to the existence and legitimacy of complex numbers. The difficulty is conceptual, and relates partly to the distinction of Platonic tradition between ideal numbers and counting numbers. The reticence about the status of the number one comes indeed from a difference of nature between the measured and the unit of measurement:

> It is obvious that one means a measure. In every case there is something else which is the subject, a quarter-tone in a musical scale, a finger or a foot or the like in magnitude, in rhythms a beat or syllable, and similarly in weight a standard defined weight. So it is in all cases, a quality for qualities, a quantity for quantities [...]. For one is not a real object in its own right. And this is only reasonable: one means a measure of some plurality, and number means a measured plurality and a plurality of measures.
>
> Aristotle [4, N 1 1087b]

There are many lessons to be learned from the Aristotelian treatment of the One. Foremost among them: Familiar mathematical usage often conceals tricky problems—for example, what exactly do we mean by a unit of measurement? Seemingly simple, the question is quickly much more delicate than it seems, especially in physics where technical and theoretical issues are superimposed on purely conceptual and philosophical issues. Let's take an example: consider a 20 cm long ruler. So it measures a centimetre 20 times. Our conventional unit of measurement, the cm, is then supposed to have a universal value and to identify with a problematic "standard centimetre". In everyday use, we trust the ruler's manufacturer and no one will ever worry about the ruler's implicit reference to a universal standard. However, the question remains: how to define an ideal unit of measurement? By reference to a prototype standard metre in metal? By reference to physical measurements (the circumference of the earth, the measurement of an atomic spectrum...): but then according to which underlying theory of the universe? Newtonian, relativistic or quantum physics? These theories, apart from depending in turn on universal constants subject to measurement uncertainties, are, as we know, unfinished and valid only to a certain degree of precision. All this ultimately renders the ideas of measure and measurement incomplete and subject to the further progress of science.

After Aristotle, a long tradition of philosophers and mathematicians thought and affirmed that "one is not a number". The reasons put forward until the sixteenth century, with Gregor Reisch (1467–1525) and Petrus Ramus (1515–1572),[2] always lead back to the Aristotelian argument: of course, the opposition of the One to the multiple is not absolute and the one, which is a measure of the few, is a kind of multiplicity (Aristotle [6, I 6 1056b]). Yet "there is good reason for the one not to be a number: a measurement is not itself measures.[3] The measure and the one are both principles" (Aristotle [4, N 1 1088a]).

The difficulties of thinking of zero as a number are probably easier to understand since, while in the case of one there is little to prevent it from ultimately participating in the concept of number, zero itself counts nothing, the "cardinal of the empty set" (the modern definition of zero) being a (late) theoretical construction rather than a natural entity. John Wallis (1616–1703) expresses himself on these difficulties as follows: "The word 'principle' has a double meaning, namely *primum quod sic* (the first that is such) and *ultimum quod non* (the last that is not such), as when we say that the movement begins at the last instant of rest or at the first instant of movement. Thus, in numbers, zero is principle in the sense of ultimum quod non and one in the sense of primum quod sic. One is therefore a number, but not zero." (Gericke [55])

Condillac discusses the subject with customary clarity:

> A word naturally becomes the sign of an idea, when that idea is analogous to the first one it signified, and then it is said to be used by extension.
>
> But, because this first idea is not always known, or because one does not know how to grasp the analogy that leads from one meaning to another, it is often regarded as an abuse to use the same word to express ideas which, though analogous, are not the same in all respects. Sometimes one is even more grossly mistaken: for, without realizing what a word means, one assumes that it always has the same meaning, and one adds absurd or childish questions.
>
> Those, for example, who have asked whether the unit is a number, have not seen that the word number has two different meanings. In the first one, it is only said of a multitude of units: and it is then obvious that the unit is not a number: it differs from it, like the simple of the compound. But, because numbers are made up of units, analogy has by extension given the simple unit the same name as several units put together, and the unit has become a number.
>
> Condillac [30, p. 428]

Condillac is right, but does not take into account the concern, evident in Aristotle, but also present throughout the history of the philosophy of number, not to reason by analogy or extension of the semantic field of concepts such as number, and to prefer instead rigorous conceptual characterizations. The reluctance to include one or zero in the field of numbers was not due to a lack of analogical audacity, but to the very concept of multiplicity and its internal logic.

The modern, set-theoretic, point of view here offers a grid of reading and interpretation complementary to that of Condillac. There is a subtle difference

---

[2]See Gericke [55].

[3]Tricot [6] translates "the unit of measurement is not a plurality of measures".

(which was identified very late in the mathematical literature and which is not always easy to grasp for mathematicians at the very beginning of their career) between the set with an element and this element itself. This difference has less to do with the element in question than with the way it is considered: either as an object, as an individual, or as an element of the set that contains only it. It underlies Aristotle's reluctance to call "one" a number. This reluctance would thus be more fundamentally due to a sub-determination of the concept of collection or multiplicity: a multiplicity (a set) is not simply a juxtaposition of individuals, of units: it is a juxtaposition of individuals seen as a totality.

In mathematical language, the difference is clearly expressed: a statement such as "let us consider the vertices A, B, C, D of a square" falls under the first point of view (juxtaposition, here in the form of an ordered sequence) while "let us consider the set $\{A, B, C, D\}$ of vertices of a square" falls under the second. The two mathematical objects (the sequence A, B, C, D and the set $\{A, B, C, D\}$) are distinct, whether one adopts a formalistic or intuitive point of view. The question "how many?" (how many vertices are there in the square?) has the effect of forcing us to consider these vertices as elements of a totality and thus to privilege ultimately the set-theoretic point of view.

Now, this question ("how many?") is legitimate, whatever the family of objects to which it is applied (I can ask "how many marbles are left in this bag?" even if there is only one marble left or none at all). Condillac's argument is finally, as we can see, less obvious and conclusive than it seemed: it may well be the internal logic of numbers that pushes us to consider one (and zero, to which we will return) as a number in its own right, and not a mere analogy or a mere linguistic artifice. All this will become much clearer with Frege and Husserl.

6.4   The Ontological Difference

Aristotle, if he knows how to think about the unity of arithmetic and geometry through the notion of measurement, also seeks to deepen the understanding of their differences, already pointed out through the opposition between discrete and continuous. This other Aristotelian attempt is important for two reasons. For mathematics, especially contemporary mathematics, it is difficult to think in a theoretical way about what constitutes a field of knowledge as such. For example, how is combinatorics defined? Topology? Analysis? On a daily basis, everyone knows how to decide what belongs to this or that discipline, but this knowledge is largely empirical and, since the twentieth century, has been confronted with permanent shifts: what Dieudonné (1906–1992) called transfers of intuition from one mathematical discipline to another.

These issues are not without practical consequences. A topologist who is asked by a young student how topology differs from geometry and analysis will, as in the Platonic dialogues, be quickly embarrassed if, after trying to define it as the discipline that studies forms making abstraction of quantities, the discussion goes on a little too long. He will first have to explain that topology is subdivided into general topology and algebraic topology, which are quite distinct areas, and that algebraic topology owes much of its ideas and techniques to group theory, ideas and techniques many of which are shared with algebraic geometry. Little by little, from analogy to analogy, the beautiful definition proposed at the beginning will prove to be more and more problematic.

The Aristotelian attempt to understand the difference between arithmetic and geometry echoes these issues by trying to define the two disciplines by the nature of their objects. Although in modern times such an "ontological" approach to the constitution of disciplines raises innumerable difficulties, it nonetheless retains a legitimacy of principle. Indeed, it probably underlies most of the judgements that mathematicians make about the disciplinary divisions that exist in practice. In any case, this attempt has had extremely important and harmful historical consequences, since it has long imposed the idea that there are intangible barriers between the different sciences, linked to the inhomogeneity of their objects.

Let us recall that the central problem of Aristotelian philosophy is that of the constitution of theoretical knowledge. This knowledge is constituted from domains of objects whose subsumption under types and genera allows the corresponding science to organize and build itself. This conception leads to the subordination of scientific domains to the underlying object domains, resulting in a strict disciplinary division. Post-Cartesian modernity would rather think of scientific divisions in terms of theoretical methods and tools whose common use from one discipline to another allows the building of as many bridges and transfers.

The main consequence of Aristotle's thesis of the incommensurability of genres in mathematics was to force the strong relationship between arithmetic and geometry to exist only as an analogy. This would be the ultimate reason for the rather strange structure that Euclid's *Elements* have for us: the same questions are treated in arithmetic and geometry without the relationships between these two approaches ever really being clarified.

6.5   Arithmetic and Geometry in Euclid

Euclid's *Elements* is the most important and surprising mathematical text ever written. Even though it was developed using earlier materials, one need only think of its date of composition (in the third century BC), its length (about 500 pages in its present format), and its content, which ranges from the systematic foundations of elementary geometry to the geometry of solids and delicate arithmetic questions, to experience a similar astonishment to that which follows Aristotle's reading.

Euclid largely follows the Aristotelian theory of science, developing an axiomatic approach to geometry, based on a set of axioms conceived as propositions that would be immediately obvious. It was not until the end of the nineteenth century that the Euclidean edifice was modified to meet the strict requirements of modern axiomatics, but these were, for the most part, marginal modifications. The interest of the Euclidean text for the philosophy of arithmetic lies in its conception of the relationships between numbers and magnitudes and, at a higher level, in its treatment of irrationality—the fact that certain geometric quantities such as the length of the diagonal D and of the side C of a square are incommensurable: one can never find a unit of measurement such that D and C are simultaneously integer multiples of it.

The Fifth Book introduces the arithmetic of magnitudes through the idea of ratio. The conceptual interest of this approach lies in the fact that Euclid, in a very general way and throughout the *Elements*, seeks to refrain systematically from fixing a unit of measurement. When the problem of choosing a unit of measurement or measure is raised, it is therefore done so specifically. These nuances correspond to one of the difficulties of founding number on magnitudes: geometrical quantities only become measurable when a measure, necessarily conventional and arbitrary, is fixed. The geometric use of arithmetic notions (irrationality of the diagonal of the square...) is thus relative to a choice of reference units. If the side is chosen as the measure, it is the diagonal of the square that is measured by an irrational number; if the diagonal is chosen, it will be the side. A physicist would say that geometric concepts are scale invariant, but not the arithmetic concepts appearing in geometry.

The Seventh Book introduces arithmetic (that is, the arithmetic of numbers) proper, and familiar definitions are given: (1) A unit is that by virtue of which every existing thing is said to be one. (2) A number is a multitude composed of units.

Some definitions do have a geometric flavour: "When two numbers having multiplied one another make a number, the one that is produced is called plane; and its sides are the numbers that have multiplied one another". However, it is obvious that the existence of a natural unit in the context of numeration introduces a strong cutoff between questions of arithmetic proper and the arithmetic of magnitudes.

After Books VII to IX, Book X returns to the theory of quantities to finally confront the problem of commensurability ("Commensurable magnitudes are those that are measured by the same measure"). It will therefore be about, for example, understanding whether, given two magnitudes, a common unit of measurement can be found for them. More subtle in this respect than many authors of modern Euclidean geometry textbooks, who set Cartesian coordinate systems from the outset, Euclid thus introduces in Book X of the *Elements* relative and intrinsic arithmetic notions, the latter being those that are independent of the choice of a unit of measurement. Thus, he does not define the rationality of the length of a line segment,[4] which would imply the arbitrary choice of a unit of measurement, but says that two segments are commensurable if the ratio of their two lengths (which is an intrinsic and a-dimensional datum) is rational—and the same goes for irrationality.

There is, clearly, throughout the Euclidean text, a profound coherence which stems from the underlying existence of a specific ontology and a systematic reflection on the idea of measurement. Renaissance thought rejected the Aristotelian legacy because of the dogmatic use of Aristotle made by medieval scholasticism,

---

[4]For the sake of readability, we deliberately modernize the Euclidean terminology.

a use that served obscurantism. Even though Galileo (1564–1642) insisted on the distinction to be made between Aristotelian work as such and its misuse in medieval writings, a lasting hostility was then established among scholars. This hostility is found in Cartesian thought, questioning traditions, with Cartesianism wiping out medieval prejudices in order to build a new science. In the context of arithmetic, this hostility translated into hostility towards the Euclidean method in the name of a renewed conception of the relationships between numbers and magnitudes.

6.6   The Cartesian Revolution

Descartes' mathematical work is part of a global project: the search for a method to direct one's mind and learn to know and master the world. Its success and its profoundly innovative character, which make René Descartes one of the few great mathematicians in history, are all the more striking and remarkable. The development of Cartesian thought, as explained in the *Discours de la Méthode* [40], was in fact conditioned as much by concrete problems (such as the Pappus problem, an open and classical problem of Euclidean geometry at the time), as by the desire to reform mathematics in such a way as to make the exercise of mathematics conform to the general method to be followed in order to "lead one's reason well".

The authenticity of this will and the largely propaedeutic function granted by Descartes to mathematics are clear from all the published texts, including the correspondence. The role still accorded in France today by "selection by mathematics" reflects this central preoccupation of Cartesianism—to use the mathematical exercise of reason to learn the correct rules of judgement and deduction since, although "reason is naturally equal in all men", "it is not enough to have a good mind, the main thing is to apply it well".

The excesses to be deplored as a result of this pedagogical choice undoubtedly stem in part from the break that has gradually been made in our "post-modern" society with Cartesian ideals. The "Cartesian will to master the world" based on the methodical use of science and engineering, which some see as the origin of many evils of the twentieth century (with the Heideggerian and post-Heideggerian idea of "machination", which thinks of man in terms of will for power and appropriation of nature) is in fact, in its initial manifestation, a will at the service of humanist ideals, first and foremost the improvement of medicine.[5] It is, in fact, rather painful to see

---

[5]"It is possible to achieve knowledge that is very useful for life. Instead of this speculative philosophy taught in schools, we can find a practical one whereby, knowing the power and actions of fire, water, air [...] and all the other bodies that surround us, we could use them similarly for all the purposes for which they are suited, and thus make us masters and possessors of Nature. This is not only to be desired for the invention of an infinity of artifices [...] but also mainly for the preservation of health [...] Having the intention of using my whole life in search of a science so necessary [...]" (Descartes [40, pp. 62–64]).

how the scientific enthusiasms of the Renaissance and the Enlightenment gave way in the twentieth century to a certain pessimism about the use of scientific reason.

In any case, the Cartesian enterprise is based on dissatisfaction, partly due to the ambivalence of the Euclidean tradition: "The scruple of the ancients to use the terms of arithmetic in geometry, who did not see their relationship clearly enough, caused much obscurity and embarrassment in the way in which they were explained" (Descartes [39, p. 306]). However, this first observation was classical at the Renaissance, and Descartes' main criticisms are more original:

> When I was younger, I had devoted a little study to logic, among philosophical matters, and to geometrical analysis and to algebra, among mathematical matters – three arts or sciences which, it seemed, ought to be able to contribute something to my design. But, on examining them I noticed that the syllogisms of logic and the greater part of the rest of its teachings serve rather for explaining to other people the things we already know [...] than for instructing us of them. [...]
>
> Then, as to the analysis of the ancients and the algebra of the moderns, besides that they extend only to extremely abstract matters and appear to have no other use, the first is always so restricted to the consideration of figures that it cannot exercise the understanding without greatly fatiguing the imagination; and in the other one is so bound down to certain rules and ciphers that it has been made a confused and obscure art which embarrasses the mind, instead of a science which cultivates it. This made me think that some other method might be sought.[6]
>
> Descartes [41, pp. 20–21]

These criticisms have universal features. Every science has the faculty to degenerate, whether in its practice or in its teaching, and the corresponding decays are largely related to the very essence of the field in question. Thus, legal disciplines or the natural sciences are undoubtedly subject to the risk of appealing more to memory than to reason (with the risks that this entails for the exercise of justice, when jurisprudence takes the place of the normative exercise and parliamentary elaboration of law). Mathematics and logic are no exception to the rule, and a regular questioning of their principles, objectives, and rules of exposure would no doubt be salutary, including today. For Descartes, this questioning takes the form of a recasting of the theory of proportions: "I thought that by this means I would borrow all the best of geometric analysis and algebra, and correct all the defects of one by the other" (Descartes [40, pp. 21–22]). The result is the birth of Cartesian geometry, where the choice of units of measurement and, in larger dimensions, of coordinate systems, effectively resolves the Greek conflict between numbers and magnitudes. It obscures the underlying philosophical problems, but mathematical efficiency is often at this price.[7]

---

[6]In the quotations of Descartes, attention should be paid to the use of words such as analysis, which is not in conformity with current usage.

[7]Cartesian geometry remains in many ways very classical: it gives an important place to a renewal of the theory of proportions and other arithmetic-geometric relations, studying for example the construction of instruments that allow one to go beyond the geometry of the ruler and the compass.

6.7 The Primacy of Arithmetic

Descartes is also interested in constructing geometrically the arithmetic operations (product, square roots...). This suggests ultimately a balanced point of view in his work between arithmetic and geometry. His successors will go further and deduce from the questioning of the duality between numbers and magnitudes a primacy of arithmetic over geometry. This primacy will culminate in the twentieth century, which will propose to reconduce the continuum and all geometry to the arithmetic of integers, a point of view well illustrated by Nicolas Bourbaki's systematic work, the *Éléments de mathématiques* [16].

Let us return for a moment to one of Descartes' successors, the "Grand Arnauld", the theologian, logician and key man of Port-Royal. The pedagogical purpose of his *New Elements of Geometry* [8], which has already been briefly mentioned, was to adapt the teaching of geometry to the Cartesian style. In this, Arnauld represents a tradition revisiting the conception of geometry in its relationship to numbers and a movement of contestation of the "Euclidean disorder" often associated with the name of Pierre de la Ramée, known as Ramus (1515–1572). The architecture of Euclid's *Elements* passes from plane geometry to the theory of ratios and introduces elementary arithmetic very late (in Chap. 7). For Arnauld, this contradicts the demand of reason to go from the simple to the complex, so much so that "the *Elements* of Euclid are so confused and blurred that, far from being able to give the mind the idea and the taste for true order, they can on the contrary only accustom it to disorder and confusion" (Arnauld [8, p. 34]). This leads to a considerable novelty: the use of categories, inherited by Arnauld from scholastic tradition "imposes the idea that geometry can only be treated as a subordinate discipline of the general theory of abstract quantities and their relationships, which must be explained before coming to any spatial consideration whatsoever".

6.8 The Problem of Infinity

This new thesis of the pre-eminence of arithmetic over geometry was to develop until the twentieth century, when it first culminated in the logical work of David Hilbert before crystallizing in the structuralist approach for which Euclidean space is the third power of the continuum, itself identified with the set of real numbers. These, in turn, are based on natural numbers by a process of extension of structures allowing the introduction of negative integers and rational numbers before passing to the continuum by a process of a general topological nature (the completion).

However, the question of the pre-eminence of number over space is not definitively closed. If, in recent times, geometers such as René Thom have pleaded for a foundation of mathematics on the intuition of the continuum, other more philosophical approaches have also militated for a questioning of the founding primacy of numbers.

One of them, to which we will limit ourselves through Frege's *Posthumous Writings* [54, pp. 276–277], is based on the problem of infinity. Elementary arithmetic, that of "kindergarten-numbers" in Gottlob Frege's terminology, is incapable of grasping the existence of an infinity of numbers, a problem that is found in axiomatic-formal approaches to numbers. For Frege, infinity is a resolutely spatiotemporal phenomenon, based on the intuition of space and time:

> [...] arithmetic cannot be based on sense perception; for if it could be so based, we should have to reconcile ourselves to the brute fact of the series of whole numbers coming to an end. [...]
>
> But here surely the position is different: that the series of whole numbers should eventually come to an end is not just false: we find the idea absurd. So an a priori mode of cognition must be involved here. But this cognition does not have to flow from purely logical principles, as I originally assumed. There is the further possibility that it has a geometrical source. Now of course the kindergarten-numbers appear to have nothing whatever to do with geometry. But this is just a defect in the kindergarten-numbers.

Frege concluded that, the more he thought about these issues, the more he became convinced that arithmetic and geometry were built on the same foundation—a geometric one. Mathematics as a whole would be geometry.

Chapter 7
Generalized Numbers I

The assertion that "one is not a number", that we encountered in Aristotle, is emblematic of a series of similar assertions punctuating the history of mathematics, and according to which this or that class of generalized numbers (zero, negatives...) would not authentically be a class of numbers.

However, the case of one is quite particular since its specificity is of a conceptual nature and stems from the qualitative distinction between measurement and unit of measurement: the problem of "one" is a problem of ontological status rather than legitimacy. The other numbers that we will consider in this section raise other difficulties since, in contrast to one, whose existence has never been questioned, these other numbers have been difficult to accept by mathematicians insofar as their very modalities of existence were poorly understood.

7.1 Zero

The history of zero is well known and we limit ourselves here to a quick review, emphasizing only those points that seem to have particular significance for the philosophical theory of number.

The very birth of zero is ambiguous: everything depends in fact on what we mean by zero and require for its conceptual extension. Its first appearance is commonly attributed to Babylonian mathematics and its descendants, particularly Greek astronomy, where it appears as a symbol in a positional numeration. The use of position numbering (such as ours) with, for example, a hundred, two tens and three units represented by the symbol 123, imposes the invention of a notation that distinguishes 1.1 (a hundred and one unit) from 11 (a ten and one unit); hence, within the numbering systems of Antiquity, the use of various symbols, including o, found in Ptolemy and used perhaps as the first letter of οὐδέν (nothing) (Gericke [55, p. 47]).

© The Editor(s) (if applicable) and The Author(s), under exclusive license to Springer Nature Switzerland AG 2020
F. Patras, *The Essence of Numbers*, Lecture Notes in Mathematics 2278,
https://doi.org/10.1007/978-3-030-56700-2_7

Speculations about ancient thought are obviously questionable when they are not based on explicitly stated theses, but it is generally accepted that this first use of zero is purely positional and does not imply the existence of a correlative idea of number. This is a very interesting theoretical configuration, which can be found in the symbolic constructions of modern mathematics where the emergence of a notion (in this case zero as a symbol within calculations) can be led back, on the one hand, to notational conventions and, on the other hand, to the very logic of calculation and operations. To a certain extent, and in opposition to a naive idealistic conception of mathematics, these were the very requirements of mathematical work (the complex calculations of Greek astronomers) and those of the signitive organization of the materials of calculation that gave rise to a new idea whose primary meaning would gradually be extended, as later civilizations realized that this symbol filling the "empty place" of the calculation of position arithmetic was susceptible to have many other meanings than those initially attributed.

This phenomenon therefore has a general meaning: at a somewhat advanced level, it is indeed common to be confronted with configurations (of calculations, symbols, proofs) where the mathematician detects both typical shapes, recurring problems, and the possibility of accounting for them by introducing new notations or new concepts. A judicious choice of notation, in particular, often makes it possible to clarify immediately what was previously obscure. The sixteenth, seventeenth and eighteenth centuries were excellent in this game, with in particular Descartes, Leibniz (1646–1716) and Euler (1707–1783) who, by creating many of the fundamental notations of modern mathematics, also established the language in which mathematics was thought.

As for zero, its history continues with India, which took it from the status of a symbol to that of a number in its own right. This evolution went through the codification of calculation rules ($a + 0 = a$, $a - a = 0$, ... (Gericke [55])). The philosophical significance of this Indian discovery, that would later spread to the Arab world before reaching Europe during the Renaissance, is however difficult to grasp. Georges Ifrah, in his *Universal History of Numbers* [75], in which an important place is reserved for India, insists on the ubiquity in Indian thought of the notions of emptiness, nothingness... which could have served as a ground for the evolution of zero from the status of symbol to that of number. The fact that Indian civilization was able to overcome the epistemological obstacle and calculate resolutely with zero, going as far as division by zero (Gericke [55]), argues in favour of such a thesis. However, "the Indian mind has always had an extraordinary inclination, flexibility and power in the calculation and handling of numbers" (Ifrah [75]). In modern times, Indian mathematics has actually given rise to the most gifted and creative mathematician in the manipulation of complex arithmetic symbols (formal power series arithmetic, in particular): Ramanujan, maybe another sign of an ability to free oneself from, or go beyond the meanings usually associated with such objects.

7.2   Phenomenology of Zero and One

Once Indo-Arabic numeration and calculation methods and zero were fully accepted in Europe, the debates on the nature of zero and even one did not die out: we still find traces of them in Frege's *Foundations of Arithmetic* [50], which devotes important developments to them.

These debates crystallize, among other things, an opposition between the Anglo-Saxon empirical tradition represented in the nineteenth century in mathematics by John-Stuart Mill (1806–1873) and the Platonic-type idealism of Gottlob Frege. If, as the empiricists maintain, numbers are derived from experience and legitimated only by it; if they are constructed from it through abstract processes, then zero and one must be inhomogeneous to other numbers since they do not refer to the experience of collections and multiplicities but to radically different experiences: that of the object as an entity and that of nothing, of emptiness. At the end of the nineteenth century, this was contrary to the daily experience of mathematicians, as the use of zero, one, negative and complex numbers had become sedimented in scientific curricula and the daily practice of arithmetic.

Frege and the idealistic current thus use this theme of one and zero to proclaim the abstract, ideal character of numbers against any attempt to bring them back to empirical foundations. Phenomenology, the philosophical theory created by Edmund Husserl, was born on this very particular ground, which is that of the theory of number at the end of the nineteenth century, where an opposition between empiricists and idealists crystallizes, but where, as we will see later, the birth of the mathematics of the twentieth century, and of set theory, is at stake.

Phenomenology, which will be discussed later in this book, is a complex theory which is both heir to the metaphysical and critical tradition (in many respects Husserl is close to Kant) and concerned with integrating the achievements of the cognitive sciences of the end of the nineteenth century (psychology, in particular). It is thus an idealistic conception of science, but where ideas and concepts are studied from the point of view of their formation from the original field of lived experience.

Husserl, in his first major philosophical work, the *Philosophy of Arithmetic* [71], taken from his habilitation thesis, is interested, among other things, in the problem of zero and one. His ideas show how to move beyond classical analyses towards a renewed understanding of the philosophy of mathematics and will serve here as a first introduction to the phenomenological method.

> Let's take the ancient definition as a starting point: number is a quantity of units. Underneath it lies the serious misunderstanding of many authors that number is a special kind of multiplicity of equal objects. Just as there are multiplicities of apples, stones, etc., so there are also multiplicities of units.
>
> Husserl [71, Chap. 8]

The result is the "very coarse" idea of unit as "absolute partial content" that Husserl finds in Locke (1632–1704), Leibniz and Berkeley (1685–1753). However:

> Any enumeration would be totally meaningless if the sign "1" or the word "one" did not have the meaning corresponding to the concept of one, that is, if it did not designate the abstract process that removes the limitation of the singular object determined from the multiplicity to be enumerated by transforming it into a simple something or one... By restricting oneself to enumerating merely as an external mechanical process, we have simply forgotten to see the logical content of thought that gives it justification and value in our entire mental life.
>
> Husserl [71, Chap. 8]

Several ideas emerge, explicitly or implicitly, from this text. The first of these is the very idea of phenomenology as Husserl formalized it a few years after the *Philosophy of Arithmetic*: the description of abstractive processes such as those allowing us to go from a collection of objects to the idea of a multiplicity of units requires, in order not to limit ourselves to rough descriptions, a specific know-how (intentional analysis in the terms of phenomenology).

Intentional analysis aims to identify, in the manifestations of our relationship to the world and to knowledge, the structures implicit in that relationship. To take a simple example, the intentional analysis of a sentence such as "I see five fruits on this table" will seek to understand the universal cognitive structures involved. Concretely, it will seek to understand the fact that there is an abstraction process underlying the passage from a sensitive experience to the subsumption of the five "things on the table" under a common concept, that of fruit, and then a subsequent passage consisting in forgetting the singularity of each of these fruits to transform them into as many "ones" and to arrive at a numerical statement.

Intentional analysis, although it may seem no different from the classical methods of the theory of knowledge, is specific in that it thematizes these phenomena by focusing specifically on the intrinsic meaning of these different passages, in the same way that mathematical logic is not directly interested in mathematical objects and theorems, but in what is universal in their rules of production.

This is the second idea underlying the Husserlian text and, beyond that, all his work: mental processes at work in everyday life and in scientific thought have a universality of their own. It is on this universality that the very possibility of theoretical knowledge is based. The insistence placed in the philosophy of mathematics, since the beginning of the twentieth century, on formal logic leads us to forget this universality aimed at by phenomenology, which is that of cognitive processes. Yet this is an essential question: no mathematician works according to purely logical mechanisms, the bulk of the creative work is done in the conceptual grasp of the objects under consideration, through analogies or through calculations, blind but guided by an implicit strategy that has led to making precisely these calculations rather than others. These are phenomena to which formal logic is blind, and which phenomenology would like to be able to account for.

7.3   Zero and One as Numbers

At work on the concepts of zero and unity, Husserl's analysis allows new horizons to emerge. After critically reviewing the definition of number as a multiplicity of units, Husserl takes a closer look at its inadequacy and at the fact that this definition more or less legitimately excludes one and zero from the domain of numbers:

> Let there be no objection that zero and one are not numbers in the same sense as two and three! The number answers the question how many, and if one asks, for example, how many satellites does this planet have, one can expect the answer to be zero and one as well as two and three, without the meaning of the question becoming different. [...] What does not fit with zero or one cannot be essential to the concept of number.
>
> Husserl [71]

However, it is necessary to look at things more closely. Any assertion must have a reason, a justification, and it is without further justification that Husserl, and others with him, put forward the almost nominalist definition of number as an answer to the question "how many?". An analogy will help to understand its limits: one can decide to call "place" any answer to the question "where?" but then one must admit that "nowhere" is a place, an idea to which common sense is repugnant. Our acceptance of "zero" as an answer to the question "how much?" has to be grounded both on grammatical reasons (relating to the structure of language in its daily use) and on formal reasons related to the internal structure of mathematical calculation, since refusing to treat zero as a number in the calculation would lead to unjustified difficulties from a practical point of view.

Intentional analysis is thus immediately led to go beyond the elementary mechanisms of constitution of the number from the phenomenon of numeration and to focus on problems of a radically different nature: the structures of language and grammar, formal and internal drivers of mathematical development; the progressive displacement of fundamental categories and definitions due to the progress of scientific thought.

"The internal reason for this situation is to be found in the similar nature of the relations that join together the numbers of the extended domain" (Husserl [71, p. 180]): the legitimacy of the extension of the domain of numbers, which begins with one and zero, but then extends indefinitely with the acceptance of negatives and then of innumerable other classes of numbers, is therefore ultimately based on a logic that is largely internal to mathematics and to the very functioning of calculation. The difficulty encountered by Husserl's phenomenology after the *Philosophy of Arithmetic* and, with it, the mathematical philosophy of the twentieth century, was to reconcile this internal and formal dimension of mathematical progress with the general theory of knowledge, based on the mechanisms of thought and cognition.

7.4 Negative Numbers

As with zero, the emergence of negative numbers and their acceptance as numbers in their own right was long and gradual.[1] Babylonian astronomical tablets, between the seventh and eighth century BC, already suggest, if not the concept of negative numbers, at least certain uses, certain notations, with an implicit idea of negativity, at least through the idea of opposition between increase and decrease. The rule of signs (minus by minus gives plus...) appeared in Diophantus in the third century,[2] but it seems that, for him, the conception of negative numbers remained operational.

The Chinese, for not having conceived arithmetic in an abstract way, developed procedures for the resolution of linear systems which corresponded more or less to calculation by determinants and to the use of conventional representations of negative numbers.

It is, once again, to India that one generally attributes the actual discovery of negatives as numbers with, among other things, the enactment of explicit rules of calculation. However, this status of number, in India as in medieval Islam, remains partly unachieved and the underlying ontology difficult to explain. The first negative solutions of equations appear in Leonardo de Pisa, known as Fibonacci (c. 1180–1250).[3] The very context of this appearance is not without interest: Leonardo de Pisa is looking at a problem of money distribution. In this situation, a negative number is interpreted as a debt: it acquires a certain legitimacy and ontological depth. Where, in the context of abstract arithmetic, the obtaining of a negative solution might stumble over questions as to its nature and mathematical admissibility, the solution obtained by Fibonacci poses little problem by inheriting from the "world of life" a concrete and immediate meaning.

In the fifteenth century, authors such as Chuquet (1484) found "abstract" negative solutions of equations, and the decisive step towards abstraction was thus taken, at least in the context of calculus. Jacques Sesiano [105] discovered occurrences of these in Provençal-speaking authors who would thus have been the first, around 1430 and 1460, to accept the idea of negative solutions without any ontological restriction, an indication, perhaps, that the high level of culture and civilization of what was to become southern France, well attested in the field of humanist culture, did not go without contributing to the development of mathematics.

As for the ontological problem—do negative numbers have a full existence, in their own right, as mathematical objects?—it will continue to animate disputes and

---

[1] See Gericke [55] and Schubring [104]. We develop here only a few aspects of the emergence of negative numbers (those that have an important conceptual dimension or involve the major actors of the epistemological debates of their time), and refer the reader to these works for a systematic historical overview.

[2] See Gericke [55] and Sesiano [105].

[3] On the history and epistemology of negative numbers, in addition to H. Gericke's work, one can consult Schubring [102, 103] and Sesiano [105].

quarrels between mathematicians and philosophers, at least until the nineteenth century. As d'Alembert wrote in the *Encyclopédie* in the eighteenth century:

> It must be admitted that it is not easy to fix the idea of negative quantities, and that some clever people have even contributed to confuse it by the inaccurate notions they have given of them [...]. Thus, negative quantities actually indicate positive quantities in the calculation, but which have been assumed in a false position [...]. There is therefore no real and absolute isolated negative quantity: −3 taken abstractly does not present any idea in mind; but if I say that a man has given another −3 crowns, it means in intelligible language that he has taken away 3 crowns from the other. [For all that], the rules of algebraic operations on negative quantities are accepted by everyone, and generally received as exact, whatever idea one attaches to these quantities.[4]

Even more than with the zero, the contrast is striking between the requirements and logic of calculation which, since antiquity, have imposed a use, albeit limited and poorly mastered, of negatives, and the psychological and ontological barriers which have for a long time obstructed their acceptance as numbers and mathematical objects in their own right:

> Of course, pairs of opposite (measurable) concepts, like credit and debit, earning and loss, property and debt, also future and past, a direction and its opposite, might provide reasonable interpretations. But the cases of a negative price, a negative tax, a negative amount of merchandise might occur as well, and could not be dismissed a priori. It is to this that the general reluctance to accept negative solutions may be attributed.
>
> Sesiano [105]

7.5   Methodological Clarifications

Two attempts at philosophical and methodological clarification, put forward in Gert Schubring's analyses [103], will take place in France in the eighteenth century, in two completely independent directions. The first is found in Fontenelle's (1657–1757) *Elements of the Geometry of Infinity* [48]. In it, Fontenelle takes up a classical justification of negatives, through the idea of opposition, but superimposes a metaphysical analysis:

> The idea of positive and negative adds to the idea of quantities that they be contrary in some way [...]. Any opposition is sufficient for the idea of positive and negative [...]. Therefore, every positive or negative quantity does not only have its numerical being, by which it is a certain number, a certain quantity, but it also has its specific being, by which it is a certain thing opposed to another. I say opposed to another, because it is only through this opposition that it takes on a specific being.

Fontenelle's thesis is fairly close to the doctrine of opposite quantities systematically developed in Germany in the eighteenth century, to which we shall return later. Fontenelle's idea seems to be to conceive negative numbers (say −3) as a number (3) to which would be added an additional quality, a "specific being" enabling it to

---

[4]Quoted by Schubring [103].

be conceived in quantified opposition to another "thing". The opposition would thus constitute the negative number as an object in its own right.

The other attempt, that occurred much later, since it was not published until 1798, much closer to the modern conception that it contributed to theorize, is the work of Condillac, the author of this *Language of Calculations* [30] who militated in favour of a largely nominalist conception of arithmetic. Condillac's look at the negatives clearly shows the advantages of such an approach when it comes to concepts for which naive ontological foundation attempts prove insufficient. Condillac develops a theory of successive abstractions, starting from the phenomenal terrain where the first empirical notion of quantities and numbers is formed. The key passage is that from quantities to abstract numbers: the nominalization of arithmetic transforms its content and allows its development to be largely autonomous with respect to intuition, the logic which is that of language and then, at a higher level, symbols replacing empirical intuition.

The existence of a logic, a specific grammar of symbols, thus allows a symbolic treatment of subtraction and negatives: "A letter preceded by the symbol + indicates an added quantity, an addition, and I call it a positive quantity [*quantité en plus*]; when it is preceded by the symbol −, I call it a negative quantity [*quantité en moins*] since it is a subtracted quantity, a subtraction." (Condillac [30, pp. 277–278])

This conclusion is profoundly modern. Although its scope may need to be narrowed somewhat, since the conceptual tools that Condillac could have used to justify his theory would not be fully developed until much later. Nevertheless, there is indeed in this idea of a grammar of mathematical language and of symbols understood as a set of rules conferring to symbols a meaning of their own, an astonishing prophetic depth for the end of the eighteenth century.

7.6 Kant and the Enlightenment

The various debates during the eighteenth century on negative numbers gave birth, in 1763, to a famous and astonishing text: Kant's *Attempt to Introduce the Concept of Negative Magnitudes into Philosophy* [77]. The general context is that of Enlightenment and an optimism of reason theorizing the progress of humanity in both the order of knowledge and politics. At the technical level, the scientific context is that of the doctrine of opposite quantities as it was developed in Germany in the middle of the eighteenth century. Gert Schubring [103] analyzes authors in this tradition, such as Abraham G. Kästner (1719–1800). The underlying idea, quite similar to Fontenelle's, is to conceive of negatives through oppositions, a negative number being a quantity to which negative status is accorded because of its opposition to another, positive quantity, these attributes of positivity and negativity being fixed in a conventional and arbitrary manner.

This conception of negativity is not without its problems: the objection that "−5 kg of flour" makes little sense outside a specific context and interpretation can hardly be evaded. However, it does have the advantage of setting the terms of a debate on the relations between numbers and quantities and their implications for the theory of knowledge in which Kant engaged in the formative years of his doctrine.

For Kant, "nothing is more detrimental to philosophy than mathematics, that is to say the imitation it makes of it in the method of thinking, where it is impossible for it to be used". Needless to say, this idea is still relevant today. However, "as regards the application of mathematics to those parts of philosophy where the knowledge of quantities is present, the situation is quite different, and the advantage is immense". (Kant [78])

In his *Essay*, Kant tries to show how negative number theory can provide philosophy with tools and original ways of thinking. Let's say it: his approach and his conclusions have, for us, very naive features. Nevertheless, the method is interesting and raises deep questions. Every mathematician knows that his mathematical knowledge is mobilized every day, both in ordinary judgment activities (for example in logical or probabilistic inference reasoning) and in the analysis of theoretical discourse—perhaps because, beyond the habit of calculation, mathematicians are naturally trained to analyze the formal structures underlying thought. It is difficult, and it would probably be erroneous, to exclude the possibility that these faculties of analysis and judgement go together with the possibility of giving conceptual and methodological tools to philosophy, even beyond the realm of epistemology and philosophy of science. This is the yardstick by which the Kantian project must be understood.

Kant follows then the doctrine of opposite quantities: "A quantity is negative in relation to another quantity inasmuch as it can only be united to it by opposition, that is to say, one suppresses in the other a quantity which is equal to it. This is certainly a relationship of opposition, and magnitudes that are opposed to each other mutually cancel out the same thing from each other, so that in reality we cannot call any magnitude absolutely negative, but we must say that, in $+a$ and $-a$, one is the negative magnitude of the other; but, as it can always be added by thought, mathematicians have got into the habit of calling negative magnitudes those which are preceded by the symbol $-$; however, we must not forget here that this designation does not indicate a particular kind of thing according to its intrinsic nature, but that relationship which unites them by an opposition to certain other things marked with the symbol $+$." (Kant [78])

Kant draws several lessons from this, we will consider here the two most original. The first is the possibility of considering a (quantitative) physics of the soul. This is a first step towards the cognitive sciences. Indeed, Kant remarks that the theory of the mind and the passions of the soul (everything that affects us, causing joy, anger, love...) can, at least programmatically, be understood in terms of the mode of opposition between the positive and the negative: "We will take an example from psychology. The question is whether displeasure is simply a lack of pleasure or a principle of deprivation of pleasure which is something positive in itself, not simply the contradictory opposite of pleasure, but what is opposed to it in a real sense, and whether, in this way, displeasure can be called a negative pleasure" (Kant [78]).

The second lesson is an elaboration of the first: it aims at understanding the overtaking of logical modes of thinking (opposition in the sense of contradiction) in the direction of mathematical modes of thinking. Kant points out that not every opposition is a logical contradiction. There is a real opposition, which is, for example, that of two forces of equal intensity but in opposite directions. This mode of opposition is much richer than logical contradiction, just as mathematics is infinitely more extended than logic, whose classical field of action is restricted to a few essentially formal and fairly elementary phenomena. In any case, there is thus a duplicity of negativity, conceivable both in the logical mode and in its arithmetic dimension.

Chapter 8
Generalized Numbers II

The history of the successive generalizations of number domains, from whole numbers to the most abstract modern algebraic theories, including rational, transcendental, complex numbers and quaternions, is undoubtedly the best known of all the history of mathematics. To go into details here would be of little interest, especially since it would take us away from the theme of natural numbers. We will therefore only retain its most significant features[1] for the philosophy of numbers, whose very content and perspectives were profoundly modified by their successive extensions.

8.1 Complex Numbers

Imaginary and complex numbers, according to the terminologies attributed respectively to Descartes and Gauss (1777–1855), appear in the sixteenth century, with in particular Cardano (1501–1576) and his *Ars Magna* of 1545 and Bombelli (1526–1572). Cardano writes for example $5.\tilde{p}.R.\tilde{m}.15$ for the sum of 5 and the square root of -15. The existence of such a notation is not insignificant. It supposes, first of all, the existence of adequate symbols to represent square roots and elementary operations on numbers. This is a recurring fact in Post-Renaissance algebra: the existence of adequate symbolisms and familiarity with these symbolisms and their uses have greatly facilitated the emergence of new computational practices and algebraic concepts. This is because, as Condillac noticed well after Cardano, algebra is the language of calculation, and this language has its own logic which can

---

[1] For a detailed history of number extensions, we refer to Gericke [55], from which most of the facts and references reported below are taken. More advanced mathematical insights can be obtained from the reading of Ebbinghaus et al. [45].

© The Editor(s) (if applicable) and The Author(s), under exclusive license to Springer Nature Switzerland AG 2020
F. Patras, *The Essence of Numbers*, Lecture Notes in Mathematics 2278,
https://doi.org/10.1007/978-3-030-56700-2_8

sometimes substitute itself for the usual discursive logic and autonomously guide mathematical thought towards new horizons.

An important part of the history of complex numbers lies in this remark: the ontology of complex numbers remained uncertain for a very long time, even though these new numbers already had in Italian arithmetic, at the end of the Renaissance, an obvious form of legitimacy due to their operative character and to the veracity of the results they allow to be obtained. In other words, one can calculate with $\sqrt{-1}$ without any ambiguity and, provided that one respects a certain number of formal rules generalizing the rule of signs,[2] one will never end up with contradictions. In modern language, we would say that complex numbers form a conservative extension of the lower domains of numbers: all identities between integers, rational or algebraic numbers, etc. that can be obtained through calculations with complex numbers are true in the domain in which they are formulated, the complex numbers ultimately appearing in this type of situation only as intermediate entities in the calculation, a bit like those intermediate figures drawn during a geometry demonstration and which are gradually erased as the demonstration progresses.

Complex numbers, admitted as tools of symbolic calculation but without autonomous existence, have long retained this status of intermediate beings. However, this was a theoretically uncomfortable situation because, beyond the *de facto* experience of the validity of these calculations, it was necessary to find a theoretical or ontological justification for them, a criterion guaranteeing this validity that was not purely empirical. Cardano was already concerned about this, a concern that would persist until the beginning of the nineteenth century.

8.2 The Fundamental Theorem of Algebra

There are many reasons why, beyond the question of its formal coherence, this or that mathematical idea moves from a marginal, problematic or uncertain status to general acceptance, or even acquires the status of an idea emblematic of a theoretical and conceptual renewal. The same remark applies to physics, where various epistemological theses aim at describing these conceptual shifts and the corresponding paradigm shifts.

In physics, the acceptance of a new theory is largely due to its ability to describe reality and to fit in with experiments: this was clearly seen at the beginning of the twenty-first century, with the expectations that surrounded the experiments conducted in Geneva, which aimed at making it possible to arbitrate between various theories (supersymmetry, Higgs boson, string theory, etc.) some of which remain to

---

[2]Denoting by $+i$ and $-i$ the two roots of -1, to "plus by plus gives plus" (*più via più fa più*) and other rules of signs, one must add rules like $+i$ by $+i$ gives -1 (*più di meno via più di meno fa meno*). These rules can be found for example in Bombelli, see e.g. Gericke [55].

this day in limbo with respect to accepted and acceptable physical theories, due to the lack of any empirical validation.

In mathematics, things are to some extent more complex. The validity of a theory does not guarantee its success, and back issues of mathematical journals abound with articles on themes that have disappeared or are moribund, due to a lack of strategic interest for the discipline once their hour of glory is over. It is tragic and sad to see a theory wither away, to see its notoriety and its interest disappear, and to see those who had developed it, when they are too advanced in their careers or unable to take up new themes, gradually fade from mathematical fame with it.

The case of complex numbers is obviously different: their place in the mathematical pantheon is eternally assured and guaranteed by their multiple roles at the most elementary and fundamental levels. As for their acceptance, Helmuth Gericke largely puts forward a historical dimension that goes far beyond their de facto computational efficiency and allows us to grasp the role that aesthetics, conceptual economy and theoretical coherence play in the construction of the mathematical corpus: "The decisive argument for the recognition of complex numbers has been the fundamental theorem of algebra: that each polynomial equation of degree n has exactly n roots." (Gericke [55, p. 65])

Already in Cardano, complex numbers appeared in the solution of polynomial equations. As for the fact that a quadratic equation can have two roots, it is already found in Baghdad in the ninth century in Alchwārizmī and in India in Mahâvira but, little by little, with Viète (1540–1603) and Descartes, the link between the degree of a polynomial and the number of its roots became clearer (Gericke [55]). Girard (1595–1632) stated it in 1629: "All the equations of algebra receive as many solutions as the denomination of the highest quantity demonstrates." "Nevertheless, solutions whose existence is impossible can still be explained." And Girard argues: "When for example [. . .] the solutions are $1, 1, -1 + \sqrt{-2}, -1 - \sqrt{-2}$. One might then ask: What is the point of these solutions that are impossible? I answer: to three things, (1) to the validity of the general rule, (2) to the knowledge that there is no other solution, (3) to this that one can find the solutions of similar equations." (Gericke [55, p. 66])

The analysis proposed by Gericke is interesting in that it thus puts forward, in the acceptance of complex numbers, purely theoretical and almost aesthetic stakes, namely the will to "make true" a statement that is false when it is stated by admitting only the real solutions (the fundamental theorem of algebra, according to which any polynomial with real coefficients has for its degree the number of its complex roots, counted with their multiplicities). This is a rather frequent strategy in the history of mathematics, because it happens quite often that a statement (a theorem) has a rather inelegant formulation in a given context, where the same statement can be formulated simply in a context that is more general or more adequate to its content. The archetypal example of this is given by Euclidean and projective geometry: projective geometry, which adds "points at infinity" to Euclidean space, in the spirit of perspective theory, offers the right framework for dealing, in a simple and elegant way, with many elementary geometric problems whose solution in the Euclidean context is laborious and rather artificial.

The process will undoubtedly seem strange to the non-mathematician, who has probably been taught a rather crude and mechanical theory of mathematics whose conceptual economy would be governed by simple (syllogistical-type) rules of inference from well-chosen axioms. That a mathematical "theorem" (false in a given context, such as the statement according to which a polynomial has as many roots as its degree, false over the reals) can "claim", "demand" by its meaning, its scope, its naturalness, that a theoretical context be constructed in which it will be true, has from this point of view something profoundly disturbing and highlights the astonishing complexity of mathematical thought in its actual work. Mathematical materials constrain thought, forcing it to take paths it had not envisaged. This coercion, which almost always accompanies truly creative work, testifies to the universal, unconditional nature of mathematics, far beyond the relativistic or societal issues that are often encountered in the epistemology and history of contemporary mathematics and which, although having a certain relevance, often forget the normative and universal dimension of mathematical thinking.

8.3 The Geometrical Foundation

The requirement of an ontological legitimation of complex numbers being further reinforced by the fundamental theorem of algebra, an argument still had to be found to establish their existence. Two paths were then to be opened, radically different both in their content and in their implications for the philosophy of arithmetic. We will come back to the second, algebraic approach, later on, dwelling now on the first in the order of historical appearance, which is geometric in nature.

The idea of a geometric representation of the complex number $x + iy$ (where x and y are real) in the form of a point of coordinates (x, y) in the Cartesian plane (i.e. with a coordinate system) gradually asserts itself at the beginning of the nineteenth century with Argand (1768–1822), Buée (1748–1826), Carnot (1753–1823), Gauss and Wessel (1745–1818). Buée remarks for example that $\sqrt{-1}$ can be interpreted as a geometrical operation.[3] This is a fundamental step, not only in the legitimization of complex numbers, which inherit from geometry all its ontological force of conviction, but, beyond that, in the understanding of the relationships between numbers and space.

Traditionally, this relationship was thought of through the idea of measurement (of length, area, volume...) but, as even for plane or spatial quantities, magnitudes were related to a unit of measurement (of area or volume), they were in the end always conceivable through anchoring number in a one-dimensional continuum.

---

[3]Let us recall, in modern notation, the underlying principle: multiplication by $i = \sqrt{-1}$ can be interpreted geometrically as the construction of a perpendicular since $i(x + iy) = -y + ix$ corresponds to the point with coordinates $(-y, x)$ and the lines joining the origin $(0, 0)$ of the plane to (x, y), respectively to $(-y, x)$ are orthogonal.

The discovery of Buée, Argand, Wessel and others is of a radically different nature: not only can complex numbers organize space as a Cartesian coordinate system does, but can also be interpreted as an operation. Calculation with complex numbers becomes a geometric calculation. Space thus confers to complexes an existence in law, and the interpretation of calculations with complex numbers as identities between geometrical operations shows that the power of these calculations goes far beyond a narrow understanding of domains of numbers. The ontological difference between number and space, arithmetic and geometry, already challenged by Cartesian geometry, is thus once again profoundly questioned.

8.4 Frege and Complex Numbers

The judgment of Leibniz, who saw in imaginaries amphibious entities between being and non-being, seems definitively outdated after the discovery of their geometric interpretation. However, the philosophical stakes underlying the existence of generalized number domains are multiple. This is evidenced by the fact that, at the end of the nineteenth century, Frege and others, such as Husserl, will return to these questions and the tacit difficulties they raise.

The text (Frege [54, pp. 68–71]) on which the following remarks are based consists of notes for a review of the *Gesammelte Abhandlungen zur Lehre vom Transfiniten* [25], a collection of three texts published by Cantor from 1886 to 1888. The purpose of Cantor's articles is to plead for the acceptance by the mathematical community of the idea of actual infinity. Faced with the objections that are traditionally addressed to the latter, Cantor develops a technical analysis which shows that these objections are based on two misunderstandings: either properties which are valid only for the finite are transposed to infinity, leading to logical paradoxes; or the metaphysical idea of infinity (inaccessible, uncontrollable) is confused with its possible mathematical realizations, so that two ontological levels (metaphysical infinity and mathematical infinity(ies)) are confused and, with them, two orders of reasons. Frege agrees with Cantor, but is more critical with respect to the thought processes at work in the mathematical component of Cantor's work, with complex numbers playing an emblematic role in his analysis.

Frege's criticism is based on an analysis of the methods for creating mathematical concepts. Like many posthumously edited Fregean texts, it is expressed in a pictorial and informal way. "Many mathematicians react to philosophical expressions [crediting them with magical properties]. I am thinking in particular here of the following: 'define' (Brahmā), 'reflect' (Vishnu), 'abstract' (Shiva). The names of the Indian gods are meant to indicate the kind of magical effects the expressions are supposed to have." In Hinduism, life, like the universe, goes through three successive phases: creation, conservation, destruction. Brahmā personifies the absolute, the creation of the world, gods and beings, Vishnu the preservation, Shiva the destruction; Frege sees corresponding analogs of those three attributes in the mathematical world.

Behind Frege's words, one must read his hostility to uncontrolled processes of conceptual creation. Mathematicians must take control of the theoretical instruments that allow them to work, and this control is most uncertain when new mathematical objects are born whose existence is not a priori guaranteed. Mathematical rigour must not be limited to checking the validity of reasoning about existing theoretical objects and tools, but must also look back to the logical origins of mathematical thought. This is the whole meaning of the Fregean work, at the origins of the modern conception of mathematics whose foundations are based on mathematical logic, the respect of certain rules making it possible, in principle, to guarantee the validity of definitions and to avoid the aporias with which mathematics was hitherto confronted.

Frege justifies his reservations about the Cantorian work on the basis of polemics about the imaginaries. He resorts to a classical stylistic figure, which has largely fallen into disuse today, by imagining a dialogue between a mathematician (M) and an interlocutor (I).

M: The symbol $\sqrt{-1}$ has the property of yielding -1 when squared. [...] It can't be arrived at by any process of sense perception; a magic incantation, called a definition, has first to be pronounced over it.

The interlocutor suggests then that the definition is actually used to stipulate that the chosen pattern of printer's ink on paper is a symbol for something having this property. However, the mathematician insists: it's a symbol, but it doesn't mean or indicate anything. It is precisely by virtue of the definition that it has this property. The interlocutor concludes:

I: Mathematicians are extraordinary people! They don't care at all about the properties that a thing actually has, but imagine being able to attribute a property to it by definition.

For Frege, all this illustrates the powers of the mathematical Brahmā, other moments of mathematical creation illustrating the powers of Vishnu and Shiva. What exactly does he mean by this? According to a tradition common at Frege's time and which will be discussed later in this chapter, the signs and symbols of algebra have a creative virtue: new mathematical objects can be defined without being correlated to "actually existing" objects. The geometrical interpretation of complex numbers certainly gives them a concrete, intuitive realization, but in this it conceals in the deep and independent meaning of geometry, the power of algebra when positing the existence of a root of -1. The twentieth century will echo this epistemological "scandal" with Lacan's psychoanalytical formula, "phallus equals root of -1", supposed to unveil the mathematical character of the hidden structures of the unconscious and of sexuality.

In any case, by sticking to its algebraic definition[4] as Frege did, $\sqrt{-1}$ has no referent other than itself and only makes sense by the rules of its use as enacted during the late Renaissance. Frege never accepted the view that mathematical beings could exist by virtue of their symbolic definition alone, independently of

---

[4]So, without referring to arguments drawn from geometry.

any reference to an intuition that would go beyond the formal intuition at work in the manipulation of signs and the rules of use that govern them. For him, the sign, the symbol must always refer to an objectivity that justifies its existence. This is one of the reasons why, despite his attachment to logic and despite the fact that he was at the origin of the formalization of early twentieth-century mathematics, Frege is frequently referred to as a representative of Platonism.

8.5   The Principle of Permanence of Formal Laws

In the 1950s, N. Bourbaki sought to theorize the overcoming, in the course of the twentieth century, of the Fregean ontology (Bourbaki [17, E IV, 52]): "The essence of mathematics—this elusive notion that had until then only been expressed under vague names such as '*reigle générale*' or '*métaphysique*'—appears as the study of the relations between objects that are no longer (voluntarily) known and described but by only a few of their properties, the very properties that we put as axioms at the basis of their theory."

In practice, however, the problem arises of deciding which axioms and relationships should be used as a basis for the theories and which rules of reasoning are allowed. In the particular case of complex numbers, in addition to geometry, algebraic arguments due, among others, to Cauchy (1789–1857), to which we will return later, guarantee the relevance of the idea of the "root of -1". This will not prevent nineteenth-century mathematicians from continuing, after the emergence of the geometric interpretation and after Cauchy, to question the modalities of extensions of number domains, in accordance with the general views expressed by Georg Cantor in 1883:

> Mathematics is, in its development, completely free and subject only to the obvious condition that its conceptions are free of contradiction and have relations fixed by definitions to previous conceptions already acquired. In particular, when new numbers are introduced, the only obligation is to give definitions such that they confer these properties and allow them to be distinguished from existing numbers. Once a number satisfies all these conditions, it can and must be considered as existing and real in mathematics. This is the basis on which rational, irrational and complex numbers must be conceived as having an existence just as real as for integers.
>
> Cantor [24]

This point of view is generally associated with the name of Hermann Hankel (1839–1873), whose "principle of permanence of formal laws", of a rather elusive content, precisely states the possibility of progressively extending the domains of numbers in such a way that the "formal laws" valid at a certain level of development of the theory are subsumed under the laws of the higher level while continuing to be valid and unchanged for the sub-domain of numbers where they were defined (Hankel [57]). The classical example is that of integer, rational, real or complex numbers, with each domain extension preserving the identities that were valid at the lower levels. Hankel thus defines mathematics as "purely intellectual, a pure theory of forms which has as its object, not the combination of quantities or their images, numbers, but things of thought to which it may correspond objects or effective relations, although such a correspondence is not necessary".[5]

As Frege notes, Hankel's theses transform the very meaning of the question of the possibility of number classes:

> A thing, a substance, which would exist autonomously outside the thinking subject and the objects with which it is associated, an autonomous principle as in the Pythagoreans: this is what number has ceased to be today. The question of its existence can therefore only be addressed to the thinking subject or to the objects grasped by thought and whose relations represent numbers. Strictly speaking, the mathematician only considers as impossible what is logically impossible, i.e. what contradicts itself. From this point of view, that impossible numbers do not have to be accepted in mathematics does not have to be demonstrated. If, on the other hand, the numbers considered are logically possible, if their concept is defined clearly and precisely and without contradiction, then this question of their existence can only concern whether or not these numbers appear in the realm of the real or in that of what is effective in intuition [...].[6]

Frege emphasizes Hankel's reference to the thinking subject, and tends to interpret it as a psychological drift, although the numbers would have "nothing to do with the nature of our being". Frege's criticism is debatable, but the problem is real: once we have lost recourse to a Platonic-type ontology where numbers exist by right, the legitimization of the existence of numbers must be led back to other principles that are difficult to dissociate from the exercise of thought, since it is in thought that concepts, whatever they may be, are born. Even when Hankel considers referring the ontology of numbers to domains of objects (one can think to the legitimization of negatives through the idea of opposition or gains and losses or, in a different register, to the geometric foundation of complex numbers), he is thus careful to speak of objects considered by thought (*gedachte Objekte*) or to refer to intuition (*Anschauung*). In Hankel's text itself, an original tension emerges, both historically and conceptually, between the domain of logic and non-contradiction (guaranteeing the possibility and legitimacy of numbers) and the domain of thought, of intuition, where, through the concrete work of the mathematician, numbers actually come into being.

Frege himself wonders about these relations between non-contradiction and existence. This aspect of his thinking is often neglected or turned to his disadvantage: he would not have fully understood the axiomatic turn taken at the end of the nineteenth century, with the radicalization of the claims of logic to found mathematics. However, these are profound questions, which are difficult to avoid if we do not want to restrict mathematics to being a vain game of language and conventions: "It is also to be challenged that the mathematician does consider impossible only that which contradicts itself. The fact that a concept is devoid of contradiction is not sufficient to conclude that an object falls under that concept. And furthermore, how can we prove that a concept does not contain a contradiction?

---

[5]Hankel [57], see also Bourbaki [17, E IV 53].
[6]H. Hankel, quoted by Frege [50, sect. 92].

It does not follow from the fact that we do not observe a contradiction that it does not exist, and the rigour of the definition does not protect against this." (Frege [50])

To a large extent, the Fregean argument prefigures the debates of the beginning of the twentieth century on the possibility of proving the non-contradiction of axiom systems, debates to which we shall return. To engage too much here on these questions would take us too far from the object of this chapter, so we will limit ourselves to underlining an idea that emerges from the Fregean analysis, relevant for the understanding of arithmetic and "generalized numbers".

It is indeed advisable to dissociate the two levels of argumentation which are, on the one hand, the general problem of the consistency of a system of axioms and, on the other hand, the day-to-day work of the mathematician. The question of the relevance of a number definition can indeed be posed in a very concrete way, in the context of solving particular problems, and the difficulties raised are then specific to the types of numbers and problems considered. Admitting negative or complex numbers, notes Frege, implies modifying the conceptual functioning of number systems. With negatives, one loses, for example, the uniqueness of the notion of root ($\sqrt{2}$ and $-\sqrt{2}$ both have square 2); with complexes, the logarithm becomes multivalued (functions such as the logarithm do not extend univocally to complex numbers, one needs the theory of coverings, Riemann surfaces...). Each system of numbers and, beyond that, each mathematical object, ultimately poses the problem of its ontological legitimacy. The example of $\log(\sqrt{-2})$ clearly shows that symbolism does not confer a rightful existence on objects and that a real work of elucidation is necessary to move from a project of conceptual extension of mathematical domains to an effective and rigorous theory: consider, for example, the extension of the definition of real functions to complex functions, a problem typical of nineteenth century mathematics. All these questions have a mathematical and historical content, but they also include a non-negligible part of epistemology: the elucidation of the concrete functioning of mathematical thought. It may be thought that it is only in this triple mode (mathematical: conceptual and technical content; historical: modalities of emergence and development of theory; epistemological: underlying mechanisms of thought) that the question of the development of mathematics must be addressed.

8.6 The Symbolic Approach

The principle of permanence of formal laws, with its resolutely abstract content, is often associated with the development of symbolic and algebraic views on complex numbers and, beyond that, on generalized number systems such as quaternions and hypercomplex systems. Frege's critical analyses that have just been reported clearly show the type of epistemological issues at stake in these developments.

Symbolist and algebraic currents have multiple ramifications, all the more so since the development of multilinear algebra, to which they are naturally attached, is not univocal. Among the main trends, we will limit ourselves here to mention successively the English algebraic school, the work of Cauchy, and the emergence of the idea of laws, the first step towards the modern notion of algebraic structure.

The idea that calculus is subject to rules, and that these rules have a logic of their own, was not new in the nineteenth century. Leibniz had already glimpsed that the simplest arithmetic identities, such as $2 + 2 = 4$, are falsely obvious and implicitly involve complex processes. The eighteenth century gradually and increasingly resolutely turned its attention to the structure of mathematical formalism and symbolism. Condillac, whose theses have already been amply commented on, seems to have played an important role in this process by influencing in particular the English school.[7] The key idea is to divert, in the study of number systems, the attention from the numbers themselves and to focus specifically on operations. In other words, and according to a process that has been repeated many times in the history of mathematics, theoretical attention shifts from a domain of objects (a domain of numbers, in this case) to a higher level, that of the rules for manipulating these objects, considered for themselves (operations on numbers).

The members of the Analytical Society, founded in 1812, such as Peacock (1791–1858), Herschel (1792–1871) and Babbage (1791–1871) "had set as their objectives the promotion and diffusion of analysis in its continental form, while emphasizing the importance of appropriate symbolism and a formal approach similar to a calculus [...]." Peacock, following the tradition inaugurated by Condillac, designated algebra as a "science of general reasoning by symbolic language", and defined symbolic algebra as "a science of symbols and their combinations, constructed upon its own rules, which may be applied to arithmetic and to all sciences by interpretation" (Scholte [101]). The principle of permanence of formal laws is obviously underlying these symbolic attempts because of the need to guarantee their validity.

8.7   Cauchy's Point of View

Augustin Cauchy is a doubly interesting figure in this movement of mutation of the notion of number systems. Several key technical ideas of this development are attributed to him, but they are accompanied by ontological and methodological concerns, amply underlined by commentators, and which make his intellectual personality all the more striking for the philosophy of numbers.[8]

Cauchy's first idea, found for example in his analysis course of 1821, is to associate the validity of calculations on complex numbers with that on the reals: after all, a complex number such as $\cos a + i \sin a$ can be seen as a pair of real numbers $(\cos a, \sin a)$, and an identity between imaginary numbers is, in the end,

---

[7]See Scholte [101].

[8]For a systematic discussion of the apparent aporias of the theory of complex numbers in Cauchy, see Dahan Dalmedico [34].

only the symbolic representation of two identities between real numbers ($a + ib = c + id$ being equivalent to $a = c, b = d$).

This point of view is doubly interesting: first of all, it is quite close to the one underlying the interpretation of a complex number as a point of the plane in a fixed coordinate system. That said, Cauchy's argument clearly shows that this geometric interpretation is not indispensable: the one-dimensional continuum is sufficient to guarantee the ontological legitimacy of complexes seen as pairs of real numbers without recourse to geometry.

The question obviously remains of justifying the rules for the multiplication of complexes. In the geometric approach, it is geometric transformations such as the plotting of perpendiculars which, by being susceptible to algebraization, give meaning to $i^2 = -1$. We owe Cauchy a purely algebraic interpretation of this identity: the idea[9] is to consider the quotient of the algebra of the real polynomials of a variable $\mathbb{R}[X]$ by the relation $X^2 + 1 = 0$. In other words, two real polynomials $P(X)$ and $Q(X)$ are said to be congruent if $P(X) - Q(X)$ is divisible by $X^2 + 1$, which we denote by $P(X) \cong Q(X)$. So we have $X^2 \cong -1$, $X^3 \cong -X$, $X^4 \cong 1$, etc. The variable X behaves exactly like a root of -1. Any polynomial $P(X)$, whatever its degree, is congruent to a single polynomial of degree 1 of the form $a + bX$ and is thus characterized by a pair of real numbers (a, b), which ensures the correspondence of congruence classes of polynomials with complex numbers.

Cauchy's argument, purely algebraic, is very profound: we find it, after him, at the basis of modern number theory and of a part of modern algebra, with notions such as "ideal numbers" characteristic of the algebra of the end of the nineteenth and the beginning of the twentieth century. Arithmetic, as a study of the properties of numbers (such as those related to primality) still relies today largely on the successive re-elaborations of this idea.

8.8 The Algebraic Approach

A third great idea emerges in the nineteenth century, still in the spirit of the program to thematize the structure of operations on generalized numbers. In various contexts (classes of quadratic forms, calculations of operators…), Gauss (1789) and French mathematicians (Legendre, 1809; Servais, 1814; Français, 1813) identified the role of notions such as commutativity, associativity and distributivity.[10]

The work of Martin Ohm (1792–1872) illustrates this basic current: after having established a list of fundamental properties of the operations of addition, subtraction, multiplication, division on numbers (I: $a + b = b + a$, II: $(a + b) + c = a + (b + c)$, III: $(a - b) + b = a$,…, VIII: $(a + b)m = am + bm$,…), he underlines

---

[9]Presented in modern terms.

[10]$a + b = b + a$, $(a + b) + c = a + (b + c)$, $(a + b)c = ac + bc$, etc., identities that are "obvious" for operations on usual numbers but that hold in much more general settings.

that these rules have their own logic, independent of the objects (i.e. variables or numbers: $a, b, c, \ldots$) to which they are applied (Ohm [88]). Thus, a notion such as addition (or, equivalently, additive law) has an existence largely independent of its embodiment in the domains of usual numbers (integers, real numbers...). The German school of algebra around Emmy Noether (1882–1935) in the 1920s and 1930s, and then Nicolas Bourbaki in France in the 1950s and 1970s, taught nothing other than the primacy of the operative rule, of "structure", over the domains of objects considered.

All these ideas, developed mainly on the occasion of the problem of the extension of the domain of numbers beyond the one-dimensional continuum, ultimately contributed greatly to the evolution of the understanding of integers. The main lessons that have been retained by the mathematical community concern, in the spirit of Condillac, the possibility of conceiving operations on numbers as having their own structure, as well as the possibility of reasoning legitimately in a symbolic way on number domains. All this finally contributed greatly to transforming the very way in which the problem of ontological foundations is posed. Since symbolism, since algebra seems to have sufficient power to give meaning to entities lacking immediate intuitive realizations, perhaps mathematics can be based entirely on them? In spite of Frege's methodological and philosophical objections, this has been a profound trend, which gained momentum in the course of the nineteenth century and was ultimately resolutely put forward, on renewed technical bases, at the beginning of the twentieth century.

Chapter 9
Cantor and Set Theory

> *The one- and multi-dimensional sets form a new derived system of ideal objects over and above the given primitive domain of objects. In fact, I believe that [the underlying] process of concept formation reveals the characteristic feature of the mathematical way of thinking. Obviously, these new objects, the sets, are different from the primitive objects; they belong to an entirely separate sphere of existence.*
> H. Weyl [109]

The next three chapters cover one of the richest periods in the history of the concept of number and, beyond that, in the entire history of mathematics. The repercussions of the debates on the nature of numbers resulting from the work of Cantor, Dedekind and Frege go far beyond arithmetic and the foundations of mathematics: a whole part of twentieth-century epistemology and philosophy was built up from Frege's analyses around the function and scope of language, its ability to describe phenomenal or theoretical reality, and finally its flexibility, which allows it to play on multiple levels of meaning.

The origins of this profound transformation of the very idea of number are twofold and complementary. There is, on the one hand, the problem of the continuum and of analysis. The whole of the nineteenth century stumbled over the difficulty of formalizing analysis, from the theory of series (convergent, divergent...) to that of functions (where notions of continuity and regularity have to face the existence of "monsters" that contradict the most elementary intuitions). It was therefore necessary to give a solid foundation to analysis, and this foundational work was partly done through the logical analysis of "what is a number"—natural, rational, real... The other major source of renewal is of a more conceptual nature and can be traced back to an effort to systematize scientific thinking, which takes many forms. A typical debate at the end of the nineteenth century (remember that this was a prosperous period for psychological analysis, leading among other things to the birth of psychoanalysis, but also to phenomenology) concerns the empirical nature of numbers: to what extent can numbers be based on intuition or, more radically, on experience? Rising up against these currents, a logician like Frege would seek to

F. Patras, *The Essence of Numbers*, Lecture Notes in Mathematics 2278, https://doi.org/10.1007/978-3-030-56700-2_9

bring the formation of numbers back to pure logic, in terms largely conditioned by the Kantian tradition.

The works of Cantor and Frege, which will be analyzed in this and the next two chapters, as well as those of Dedekind, which will be analyzed later, are essentially contemporary and have the remarkable feature of being both close in their technical content and different in their theoretical and philosophical motivations. Cantor, initially motivated by problems of Fourier analysis, rather quickly oriented his work and thought towards the problem of actual infinity. Dedekind is the most algebraic minded of the three and, even if his work on the foundations of the concepts of natural and real numbers is motivated by the desire to establish analysis on rigorous foundations, an algebraic orientation is evident in the general development of his thought. Frege, more philosophical, will orient his research in a confrontation with Kantism and towards the idea of a priori analyticity.

9.1   The Path of Analysis

A quotation from Hermann Weyl, dated November 1917, taken from the Preface to his work on the continuum [109] shows to what extent the uncertainties on the foundations of the theory of real numbers have been able to last.

> It is not the purpose of this book to cover the "firm rock" on which the house of analysis is built with a fake wooden structure of formalism. Rather, I shall show that this house is to a large degree built on sand. [...] At the centre of my reflections stands the conceptual problem posed by the continuum – a problem which ought to bear the name of Pythagoras, and which we currently attempt to solve by means of the arithmetical theory of irrational numbers.

In 1932, the same Weyl would observe that his work had been overtaken by Intuitionism and Formalism. "Still this deeper grounding of the foundation has not led to an even moderately satisfying or defensible conclusion; things remain in a state of flux".

It is in this enduring context of "crisis of analysis" that Cantor's first works are born, some 50 years before the writing of Weyl's text. Georg Cantor, born in 1845 in St. Petersburg, died in 1918 in the psychiatric hospital of Halle, where he taught at the University from 1869 onwards after studying in Berlin with some of the greatest German mathematicians of the time: Kummer (1810–1893), Weierstrass (1815–1897), Kronecker. Cantor is one of the key figures of mathematics, of a multiform celebrity, due in part to the emblematic nature of the questions he tackled with unequalled originality, including the problem of actual infinity. Another important component of this celebrity relates to his intellectual personality and his motivations, some of them mystical.

His early work focused on analysis and, more specifically, on trigonometric series. The theory of series studies, in a general way, the properties of series of numbers (usually real or complex) or of series of functions. A typical problem is that of convergence: is a sequence of numbers, for example that obtained by summing all fractions $1, \frac{1}{2}, \frac{1}{4}, \frac{1}{8}, \ldots$ obtained from 1 by successive divisions by 2, convergent? Euler was one of the first architects of the theory in the eighteenth century. Very early on in the mathematical literature, we find approaches, apparently rather paradoxical, allowing for example the summation of divergent series. The nineteenth century was confronted acutely with these questions, which led mathematicians to systematically explore the properties underlying the definition of real numbers and the corresponding topological[1] notions. Trigonometric series are those series constructed from trigonometric functions (sine, cosine, tangent functions...). Their study led Cantor [20] to reflect on the one-dimensional continuum (also called the real line). He then chose to characterize real numbers by Cauchy series: any real number is the limit of a sequence of rational numbers $a(1), \ldots, a(n), \ldots$ where $|a(n + m) - a(n)|$ becomes infinitely small uniformly in m for n sufficiently large.

9.2 Measuring Infinity

Cantorian thought evolved rapidly from this early work. As far as the understanding of the concept of natural number is concerned, he made decisive but indirect contributions to it. Cantor was little interested in whole numbers, which he considered to be well understood, but he raised the totality of natural numbers and their cardinality to the status of a mathematical object in its own right. An article of 1874, "On a property of the set of real algebraic numbers", already introduces a whole set of profoundly innovative ideas—mathematically and by their epistemological correlates (Cantor [21]). We know, since the Greeks, that in addition to rational numbers (quotients of integers), there are irrational numbers, such as $\sqrt{2}$. The latter is said to be algebraic: it is the solution of a polynomial equation with integer coefficients ($x^2 - 2 = 0$). Finally, there is the set of real numbers: all the numbers of the one-dimensional continuum, which admit (in the standard model of the continuum) several definitions equivalent to that by the Cauchy series.

There are thus sets of numbers nested one inside the other, "bigger and bigger":[2] integers, rational numbers, algebraic numbers, real numbers. Cantor, in the article of 1874, establishes an astonishing property: the set of algebraic numbers is in bijection with that of positive integers, that is to say that one can, with ordinary positive integers, number all algebraic numbers univocally![3] If we think in terms of traditional mereology (the intuitive notions of inclusion, of "bigger", "smaller", etc.), this is something deeply disturbing: a set of numbers (algebraic numbers) is

---

[1] Topology is the study of the qualitative properties of space: continuity, closure, compactness....

[2] We will say that one set is "bigger" than another when it strictly contains it, reserving the terminology of "larger" for the comparison of "numbers of elements". This last notion may seem ambiguous, but it is precisely to Cantor (among others) that we owe its clarification—until we get there, we will use these notions intuitively for the time being by putting them in quotation marks.

[3] To any algebraic number corresponds a single positive integer, and vice versa.

known, since the Greeks, to be much "bigger" than another (that of integers), and yet one cannot mathematically claim that one is "larger" than the other since they have "as many" elements.

Similar phenomena, intrinsically associated with the properties of infinite sets, were already known, and before Cantor much simpler examples were known to exist where, from the inclusion of one set of numbers in the other, it is impossible to conclude that one is intrinsically "larger". There is, for example, nothing shocking or paradoxical in noting that the set of integers strictly greater than 1 is not intrinsically larger than the set of integers greater than or equal to 1: one can very well choose to number a finite or infinite sequence of objects (for example the internal pages of a notebook) starting with the number 2 rather than with 1 (for example if one wishes to number 1 the inside front cover on which one has started to take notes). It is also easy to show that the integers can be put in bijection with the positive integers. That apparently as inhomogeneous sets as natural and algebraic numbers share the same property is much more surprising, and typical of the scope of the Cantorian ideas, from which many of the most important results of modern mathematical logic originate. There is indeed a qualitative leap with this use of the infinity of integers: a technical and epistemological break.

The same article goes even further and shows that the analogous property is not true for real numbers: the one-dimensional continuum cannot be put in bijection with the positive integers. To some extent this result is reassuring: it confirms the heterogeneity to the continuum of the discrete character of integers. But its epistemological scope goes far beyond this, since Cantor establishes by this, in a rigorous, mathematical way, that the Cartesian fusion of algebra and geometry is more complex than it seems. The continuum is not algebraic: there are "many more" points on the geometric line than there are algebraic numbers. This justifies to a certain extent Euclid's embarrassment in the joint treatment of numbers, magnitudes and ruler and compass extensions of domains of numbers (in modern language, successive quadratic extensions of the field of rational numbers).

It is by the yardstick of Greek thought and the Pythagorean discovery of the incommensurability of magnitudes that the scope of the Cantorian result can best be measured. It has multiple and profound consequences. Thus, as within an arbitrary formal system one can explicitly construct only a countable number of real numbers, almost all real numbers have, from a mathematical point of view, only a virtual existence. We can certainly talk about them, and modern mathematics does not hesitate to argue about the totality of real numbers, but their individual existence remains evanescent, for lack of constructibility, and is only guaranteed by the artifices of formal logic and the problematic postulates of set theory, which will soon be discussed.

9.3 The Diagonal Argument

Cantor's first proof of this result was topological and is based on geometrical arguments such as the behaviour of nested segments. In 1891, he simplified his proof and introduced, at the same time, the "diagonal argument", an essential idea whose scope went far beyond its initial use in the 1891 article.

Mathematical progress is a complex phenomenon, and combines different conceptual moments that history does not always manage to deal with adequately. Great results are part of it—thus recently, demonstrations of Poincaré's conjecture or Fermat's last theorem. However, in the long term, major advances are often the result of less spectacular discoveries, but which lead to the emergence of new ways of thinking, new techniques or new objects and problems. The Cartesian turning point, if associated with a precise mathematical result (the Pappus problem, solved in all generality by Descartes and an emblematic illustration of his method), is thus worthy first of all as a methodological turning point.

The diagonal argument deserves to be approached from this point of view. It certainly gives an answer to one of the greatest mathematical questions (that of the relationship between the continuous and the discrete), but is also valid as a technique or universal idea: the type of idea that a mathematician, once he has encountered it in a context and appropriated it, will know how to reuse in other contexts, for other proofs, other objects. It illustrates the fact that constituted mathematics is not limited to a corpus and a set of open questions and conjectures, but has perhaps its most fascinating part in the capital of accumulated conceptual schemes. The diagonal argument, which Gödel used later to study the limits of axiomatic systems, has the advantage of a certain simplicity, but there are very many conceptual schemes of similar power in the different fields of mathematics. Here we mean scheme in a non-technical sense: a mixture of method and intuition, or, more precisely, the articulation of a method to intuitions, something that makes it possible to understand immediately the essence of a proof without having to read the details, and sometimes simply from the statement of the result.

The Cantorian evidence itself is worthy of note. Let's consider the numbers of the one-dimensional continuum (real numbers, defined for example with Cauchy's sequences) comprised strictly between 0 and 1 (this set—which we denote by I—is easily put, in a multitude of ways, in explicit bijection with the set of all reals). These numbers can always be written in their decimal form[4] ($1/3 = 0.3333\ldots, \sqrt{2}-1 = 0.414\ldots$). If I was in bijection with the natural numbers, we could order its elements in an infinite sequence: $x(1), x(2), x(3), \ldots$ and so we could equivalently order the elements of the set of all the reals. Let us write $x(i)$ in decimal form: $x(i) = 0.x(i; 1)x(i; 2), \ldots$, where $x(i; n)$ denotes the n-th term of the decimal expansion of $x(i)$. For example, if $x(3) = \sqrt{2} - 1$, $x(3; 1) = 4$, $x(3; 2) = 1$, $x(3; 3) = 4, \ldots$ Let us then consider the number y with decimal expansion

---

[4]To be precise, one has to take into account the fact that some real numbers have two decimal expansions.

$y = 0.y(1)y(2)\ldots$ with $y(i) := 1$ if $x(i; i) \neq 1$ and $y(i) := 2$ otherwise. The number y thus constructed belongs to the one-dimensional continuum, is between 0 and 1 but, still by construction, is different from all $x(i)$ since it differs from each of them by at least one term of its decimal development. Hence the impossibility of the existence of a bijection between the set of real numbers and the set of natural numbers.

9.4 Set Theory

The next step in the evolution of Cantorian thought is the development of an abstract theory of sets. In 1878, he published a "Contribution to the Study of the Theory of Multiplicities" [22]. The language of sets was not fixed at the time, and several competing terminologies shared the field for the notion of "set" as we use it today. German is the language in which most of these debates took place, but this variety of terminology can be fairly well captured by the words "multiplicity, collection, group, set" which, by the way, intuitively correspond to different ways of understanding the notion of "collections of objects". Let us quote Cantor:

> Let M and N be two well-defined multiplicities, if M and N can be put in correspondence in a univocal, complete, element by element, way, we say that they have the same power or are equivalent.
> When the multiplicities are finite, i.e. when they are made up of a finite number of elements, then the concept of power corresponds, as can easily be seen, to that of numbers of elements and, consequently, to that of a positive integer because, in fact, two multiplicities have the same power if and only if the number of their elements is the same.
>
> Cantor [22]

The remark is crucial, in that it refers the concept of positive integer to a purely set-theoretic phenomenon. It is not entirely new (Gericke [55]): other authors had, before Cantor and Frege, understood the existence of a relation of principle between the concept of number and equinumericity (the fact of being able to put in bijection the elements of two sets), but the end of the nineteenth century succeeded in bringing out its fundamental meaning. However, Cantor did not take the step of identifying the concept of number with that of power: in 1878, he cautiously remained with the idea of an "easy" correspondence. It must be said that, unlike Frege shortly after him, Cantor was mainly interested in the problem of mathematical infinity and his remarks on natural numbers were essentially incidental remarks. By the way, infinity inherits with him characteristic properties already observed by Bolzano and Dedekind: "A part of a finite multiplicity always has a power smaller than that of the multiplicity itself; these properties cease to be true for infinite multiplicities ". (Cantor [22])

9.5 The Concept of Set

At the philosophical level, Cantor remains faithful in substance to the idea of mathematics in the Aristotelian tradition: fundamental concepts, such as multiplicity, can be described, explained, but do not have to be defined. In a text of 1882, he clarifies his conception of the concept of multiplicity (*Mannigfaltigkeit*), which until then had remained largely non-thematized:

> The concept of power can be seen as the most general and pure moment of multiplicities. It is an attribute of all well-defined multiplicities, whatever conceptual properties their elements may have.
>
> I say that a multiplicity of elements that belong to any conceptual sphere is well-defined if, on the basis of its definition and as a consequence of the logical principle of the excluded third, it must be considered as internally determined *both* whether any object of the conceptual sphere considered is or is not an element of the multiplicity *and* whether two objects of the multiplicity must be considered as identical in spite of formal differences in the way they are given.
>
> Cantor [23]

This seemingly innocuous text raises several philosophical problems that are not without interesting purely mathematical correlates. The idea of conceptual sphere is thus far from being neutral. Whatever may have been Cantor's intention in this case, the text suggests that a multiplicity is only conceivable with reference to a horizon. This corresponds to our intuitive use of multiplicities and numbers and, more generally, of language, since all our judgements and statements make sense in a specific, albeit often implicit, context. It also suggests to us that mathematical objects always exist against the background of an implicit universe such as the Euclidean plane, the space-time of relativity, the theory of groups. . . So that a silly question like "Do the elements of the algebra of quaternions belong to the Euclidean line?" is, as it stands, simply meaningless. But we will see that, in Frege's case, for whom the conception of numbers is a matter of logic rather than of mathematics, such a question would be legitimate. This raises problems, since the paradoxes of set theory that ruined Frege's ambitions arose precisely from the abandonment of the "normal" rules of mathematical language usage in favour of purely syntactic rules.

The mathematical anchoring of Cantorian set theory is still attested by the idea of definiteness:

> In general, the decisions corresponding [to the determination of whether or not an element belongs to a set] cannot be made with the help of existing methods in a sure and certain way; but this is not what it is about, but only the *internal determination* which, in concrete cases where the ends require it, can be performed up to an *effective external determination*.

The explanation of these rather obscure considerations follows: they relate to the usually very difficult problem of deciding whether a number is algebraic (i.e. solution of a polynomial equation with integer coefficients) or transcendental. "I recall as an example the case of the set of algebraic numbers, which can be designed in such a way that membership in the set of algebraic numbers of a given number is determined internally."

Here "internal" is to be understood as: according to the nature, the very definition of the set under consideration. For Cantor, it is natural to think that a real number must be algebraic or transcendental, even though there would be no way to decide this at a given stage of mathematical development. However, these questions are difficult if one prohibits the use of the excluded third, whose legitimacy is far from self-evident when dealing with this type of problem. Cantor could not have been aware of this: indeed, this type of logical questioning on the functioning of mathematical thinking would be one of the consequences of the emergence of set theory.

9.6 The Legitimacy of Infinity

A text from 1883 [24] marks another step in the Cantorian philosophy of number. In it, Cantor wondered about the reasons for the introduction of infinity in mathematics: "We can consider numbers to be actual (effective) insofar as, on the basis of definitions, they occupy a perfectly determined place in our faculty of knowledge, are clearly distinct from the other constituents of our thought, and thus modify, in a very precise manner, the substance of our intellect." And elsewhere: "We can attribute to them actuality (effectiveness) in so far as they are to be regarded as an expression or image of worldly processes or relations, worldly being understood as opposed to intellectual."

Looking for foundations for the use of infinity, Cantor revives thus the Aristotelian problem of the separation of mathematical concepts from the "real world". These ideas, and in particular that of the effectiveness of mathematical concepts, have a strong ontological significance. That effectiveness in the first sense (internal to the intellect) cannot go without the second (mundane) does not go without evoking Kantism, according to which the structures underlying mathematical thought reflect the structures of space-time, i.e. of the ambient world. The accent, the tone of Cantorian thought is, however, unlike Frege's, quite distant from Kantism and manifests a true originality that does not go without aporias. Let us recall indeed that this "actualism", this Cantorian realism goes hand in hand with the legitimization of formal practices since "mathematics is, in its development, completely free and subject only to the obvious condition that its conceptions be free of contradictions..."!

9.7 The Activity of Thought

His latest texts confirm the singularity of his thinking and his malleability, which allow him to reconcile formalist, metaphysical, logical and intuitive influences in mathematics. This protean and aporetic character makes, with the depth of his mathematical advances, his philosophical and historical interest: Cantor is the man of a period of transition, he reflects its complexity, uncertainties and genius.

In a text of 1895, he reaffirms the anchoring of numeration in the activity of thought: several classical theses that have emerged since Greek Antiquity can be recognized.

> By a set, we must understand any grouping into a whole M of definite and separate objects m from our intuition or our thought.
>
> Every set has a definite "power", which we also call its "cardinal number". We call by the name "power" or "cardinal number" of M the general concept which, by our active faculty of thought, arises from the set M when we make abstraction of the nature of its elements m and the order in which they are given to us.
>
> We denote the result of this double act of abstraction, the cardinal number or the power of M, by $\overline{\overline{M}}$. Since any element m, if we abstract from its nature, becomes a "unit", the cardinal number $\overline{\overline{M}}$ is a defined set of units, and this number exists in our mind as an intellectual image or projection of the starting set M.
>
> Cantor [26]

Cantor then takes up again the definition of equivalence given in 1878. In the meantime he decided to identify the concepts of power and cardinal number and concludes: "The equivalence of the sets thus forms the necessary and sufficient condition for the equality of their cardinal numbers."

Following, as we have done, the movement of Cantorian thought in the detail of its development, we must conclude that the concept of equivalence is the logical moment on which the mathematical concept of cardinal number is based. The "psychological" description corresponds to the metaphysical moment on which it is based; the two moments come together in the unity of a mathematical knowledge that has a mundane effectiveness and allows calculation, even with infinite numbers.

Chapter 10
Frege's Logicism

Frege is the author who has contributed in the most profound, original and decisive way to the understanding of numbers and the relationship between them and the springs of theoretical thought. This is because the Fregean project, unlike the Cantorian project, is not mathematical; at no time does Frege seek primarily to demonstrate new mathematical statements. What interests him first of all is what is the subject of this book: the pure laws of thought and the emergence of the concept of number. However, Frege has a mathematical background, and his thinking was always marked by a certain form of realism or platonism that is very common in the mathematical community.

The scope of his work is considerable, since it underlies the foundations of modern mathematical logic. The introduction to the translation of his first great text, the *Begriffschrift* [49], in the reference volume gathering the founding texts of mathematical logic (van Heijenoort [59]) leaves no doubt about the importance of the Fregean contribution. "[The *Begriffschrift*] is the first work that Frege wrote in the field of logic and, although a mere booklet of eighty-eight pages, it is perhaps the most important single work ever written in logic." Its fundamental contributions, that we list here in the order of relevance for our later analysis of numbers in Frege, include the analysis of propositions in terms of the duality function/argument rather than subject/predicate, a logical definition of the notion of mathematical sequence, truth-functional propositional calculus, the theory of quantification and a logical system where derivations are conducted exclusively according to the form of expressions. "Any single one of these achievements would suffice to secure the book a permanent place in a logician's library."

This inscription of Frege in the history of modern logic, as well as his role as founding father of the so-called "analytical" logical tradition (as opposed, among others, to the Kantian tradition where synthesis plays a decisive role), have for a long time largely influenced the way in which his thought was understood. His main guiding idea, at the methodological level, is already contained in the *Begriffschrift* and in Frege's exposition of the idea of induction there: "We see how pure thought,

F. Patras, *The Essence of Numbers*, Lecture Notes in Mathematics 2278,
https://doi.org/10.1007/978-3-030-56700-2_10

independently of any content given by the senses or even of any content given by
an a priori intuition, can, on the basis only of the content that results from its own
constitution, lead to judgements that, at first glance, seem possible only on the basis
of some intuition."

Is the internal structure, the very constitution of pure thought, capable, by
itself, of founding number? Frege will not cease to try to answer this question,
which, pushed to its very end, poses, at a level of radicality unequalled and
properly unimaginable before him, the problem of the logical and intuitive roots
of mathematical thought and the nature of its contents.

Let us summarize[1] first the main lines of his intellectual journey with regard to
arithmetic and his attempt to bring it back entirely to logic—the logicist program,
that was developed further by Russell. In his 1884 *Foundations of Arithmetic* [50],
Frege criticizes the psychology of mathematicians who, for lack of rigour, give a
confusing definition of number. His own definition of numbers as extensions of
concepts, we will see later what this means, distinguishes him from both algebraists
and empiricists. Later, in the *Fundamental Laws of Arithmetic* [52] of 1893 (t.1)
and 1903 (t.2), he opposes also formalists, who would like to reduce arithmetic and
the whole of mathematics to a game with signs. Frege's version of logicism, this
attempt to found arithmetic on logic, was however stopped by Russell's paradox,
formulated in a 1902 letter to Frege. We have already said that it led the later Frege
to think that mathematics should look for geometrical foundations instead. We will
however focus here on his first ideas, more original, and especially his wonderful
contribution to the philosophy of numbers and of mathematics, the *Foundations of
Arithmetic*.

When we look closely at it, the Fregean thought is irreducible and deeply
inhomogeneous to the different philosophies of twentieth-century mathematics
(formalism, mathematical logic, intuitionism, phenomenology, structuralism...),
which makes any interpretation difficult. The key point is its profoundly original will
to produce thought contents with pure logic. In a tradition of thought that begins with
Greek philosophy and ends in part with Frege, pure logic is the science of the form
of thought: it has nothing to say and can say nothing about contents. Frege's attempt
challenged this conception of the nature of logic and, as a result, twentieth-century
mathematics was to be largely legitimized by recourse to mathematical logic. After
Frege and Russell, the Vienna Circle, Carnap, Wittgenstein, logical atomism and
later philosophies of language questioned the links between the logical structures of
language and the possibility of producing scientific statements about the world. In
another direction, phenomenology tried to justify the idea of a transcendental logic
that would reconcile the traditional formal purity of logic with its articulation to the
formal structures of the phenomenal world. They all relate ultimately to Frege, but
none of them have been faithful to the Fregean logicist program, a subtle balance of
mathematical technicity, logical analysis and post-Kantian philosophy.

---

[1] We follow here Frege's biographical notice in Lecourt [81].

10.1 The "Platonism" of Frege

The most difficult thing to understand about Frege is undoubtedly the juxtaposition of a renewal of arithmetic and his systematic opposition to formal arguments in which mathematics does not refer to a "domain of objects". This latter view has led to seeing him as one of the last great historical representatives of mathematical Platonism. His opposition manifests itself essentially in the discussion of what is and should be a definition or a system of axioms (when it is therefore a question of "creating" mathematical objects or a domain of mathematical objects):

> Most mathematicians are satisfied in researches about definitions when they have verified necessary conditions. When a definition arises naturally in the course of proof, when contradictions never arise, when relationships between apparently distinct things are recognized, and when a higher order and law of mathematics emerges, the mathematician considers the definition to be sufficiently certain and has little doubt about its logical justification. But it is necessary to be careful that the strength of a proof remains illusory, even if the chain of deductions would be flawless, as long as the definitions remain justified only afterwards by the fact that a contradiction has not been reached.
>
> Frege [50, p. 23]

In other words, the absence of contradiction is not sufficient to found the existence of mathematical objects, to justify their definition. However, where a mathematical philosophy such as Kantism or phenomenology would seek, in order to justify their existence, to produce these objects in pure intuition, Frege aimed at constructing them according to the rules of pure thought—a project that, as we know, he abandoned at the end of his career. This "logical platonism" of Frege reflects one of the three main principles that governed his writing of the *Foundations of Arithmetic*: "One must clearly separate the psychological from the logical, the subjective from the objective." (Frege [50, Einleitung X])

10.2 Frege's Relationalism

The second principle can be understood as a relativism, a contextualization of meaning. "We must seek the meaning of words in the context of the propositions, and not by isolating them": meaning is born from the relations that manifest themselves within the propositions.

The logic behind this second principle is difficult to grasp at first glance, but could well characterize Frege's Platonism in its opposition to a naive Platonism where the question of the status of mathematical objects is never really seriously debated. What he opposes is nothing less than the idea that a mathematical object, say "the triangle", exists "in itself", regardless of its mode of constitution in theoretical thought. Virtually all philosophers of mathematics have noted that we do not have a full, proper intuition of mathematical objects and concepts: what we have access to through thought and intuition are not "objects in themselves", but rules of use and recognition (schemas, in Kantian language). Frege goes further and insists on the rooting of these rules in the structures of language and enunciation.

10.3   Concept and Object

The third and last principle is the most fundamental and delicate: "The distinction between object and concept must be kept present." Frege points out that many of the apparent difficulties in mathematical philosophy are due to the oblivion of this distinction. Certainly, in ordinary language, we are most often able to make the distinction without thinking about it: it is the concept of the horse that is implied in the sentence "the horse is a quadruped", and the "object-horse" (an individual) in the sentence "the horse is tied up in the meadow". But such ambiguities are unacceptable in rigorous scientific language. And what about the number 5: is it an object, a concept? Frege explains himself in an 1892 text, "Concept and Object":

> The term "concept" has various uses; it is sometimes taken in the psychological sense, sometimes in the logical sense, and perhaps also in a confused meaning that mixes the two. For my part, I have chosen to adhere strictly to the logical use of the term. [Several criticisms having been made of the use of the word concept in the *Foundations of Arithmetic*], I will first notice that the explanations I have given there have not been proposed in my opinion as a true definition. One cannot ask that everything be defined, any more than one can ask a chemist to analyze all matter. What is simple cannot be analyzed, and what is logically simple cannot be truly defined.
>
> Frege [51]

This discussion, which applies to primitive logical notions such as "concept", does not apply to the fundamental concepts of arithmetic which, for Frege, are not logically simple and which he will try to construct from notions such as those of object, concept or relation.

So what is a concept, after all? The nature of a concept is to be predicative: what in one statement is asserted of another term. "Conversely, an object name, a proper noun, cannot be used as a grammatical predicate at all." Frege's mathematical logic is therefore functional, in the sense that a function f is always unsaturated: it is what is always predicated on something—the indeterminate variable x in the notation $f(x)$ for the function f. But Frege doesn't stop there: where the notion of function is usually used, in mathematics, to denote a correspondence (for example, $\cos(x)$ denotes the function cosine, from the set of reals to the interval $[-1, 1]$), Frege uses the idea of function to codify the logic of predicates and truth values. Thus, the unsaturated statement: "The celestial object . . . is a star" can be understood as a function having as value true when the empty space is filled by "sun" and false when the empty space is filled by "moon". His various writings largely use this atypical and true/false-valued case of the concept of function.

Yet Frege must struggle to justify his approach: after all, don't we say "This man is Alexander the Great" or "This number is the number 4" (where the predicate seems to be a proper name, the name of an object: Alexander, the number 4) as we say "This animal is a mammal" (where the predicate, mammal, is clearly of a general and conceptual nature)?

We are dealing here with one of the ambiguities of language, including mathematical language. In the first two usages, "is" expresses an identity and has the role of the arithmetic sign of equality, while in the last one we are dealing with a lexical form of attribution (this animal falls under the concept of mammal).

In a more subtle way, Frege's attempt leads him to specify a fundamental distinction, evident in twentieth century mathematics, whose emergence is associated precisely with the constitution of set theory. Frege, because of the philosophical character of his approach to set theory, is undoubtedly the one who conceived and understood it most acutely. Let's come back to the object/concept distinction: one of the theoretical difficulties lies in the fact that every object is associated with a concept. In ordinary life: this sheet of paper on which I am writing is an object in its own right, but I can very well form the concept of "object on which I am writing", a concept that I can predicate, in the Fregean fashion, from this sheet of paper in front of me, and from it alone. The link with arithmetic, not very obvious at first glance, is nevertheless very real and profound. We recognize here one of the forms of the aporia of the third man: to any object X we can associate the idea of object X (and then the idea of the idea of $X \ldots$). In mathematics, this distinction is translated into current notations by $x \neq \{x\}$: the object x, whatever it is, is distinct from the one-element set containing x.

Here again, we can see how Fregean thought succeeds in resolving, by the imposed use of syntactic and logical constraints, an aporia on which mathematical philosophy had so long stumbled: the status of one as a number. In mathematics prior to formalized set theory, there was little need to distinguish between x and $\{x\}$, and the structural ambiguities of non-formalized language could easily be accommodated. For the same reason, the distinction between an object and the same object conceived as a collection, a multiplicity, was difficult to conceive: one was not a number in its own right!

10.4   A Priori Analysis and Synthesis

Frege's desire to bring arithmetic back to logic has an explicit philosophical foundation. We have already pointed out two features of Fregeism: a renewed form of Platonism and a certain post-Kantism where, by post-Kantism, we do not mean a more or less faithful follow-up or interpretation of Kant as can be found at the beginning of the twentieth century with the Marburg school (Cohen (1842–1918), Cassirer...), but a critical attitude and sometimes even an opposition to Kantism, from which Frege nevertheless inherits a certain number of structural themes and concepts.

The most important of these is the opposition between the synthetic a priori and the analytical character of mathematical truths. For Kant, true to tradition, analysis (in the sense of formal deduction) and more generally logic are incapable of creating thought content. Mathematics in general, and arithmetic in particular, can therefore only acquire content through an extra-logical process: a priori synthesis, which for Kant is basically, in mathematics, the ability to produce thought content that codifies the structures of our relationship to the world and the very possibility of thinking, in a theoretical way, about space and time.

Kant's epistemology, apart from being limited by an understanding of science that necessarily depends on the knowledge of his time (unlike Descartes, Leibniz, Bolzano and Frege, Kant is not a mathematician-philosopher and therefore does not have the same type of access to scientific content), is dependent on patterns of thought about mathematics and physics that have been widely questioned from the nineteenth century onwards. Thus, the problem of founding mathematical analysis (the theory of real and complex series, of the functions of a variable...), characteristic of the nineteenth century, is foreign to the horizon of Kantian thought even though it conditions the research of a Cantor or a Frege. It is therefore not surprising that Frege, in spite of a deep understanding of Kantism and a certain empathy with it, seeks to renew its terms:

> There are also philosophical reasons why I have undertaken this research [about the foundations of arithmetic]. The question of the a priori or a posteriori, synthetic or analytical nature of mathematical truths awaits its answer here. For, even if these concepts (of apriority, analyticity...) are philosophical, I believe that the decision cannot be obtained without the help of mathematics.
>
> In truth, it all depends on the meaning we give to these questions. It is not uncommon that one first obtains the content of a statement by one method, and only then, by another, more difficult method, does one provide rigorous proof, through which one can often specify more precisely the conditions of validity of the statement. In general, a distinction must therefore be made between the way in which we arrive at the content of a judgment and the way in which we justify its assertion.
>
> Frege [50]

In other words, is it necessary to distinguish the mode of constitution (necessarily empirical and psychological) of the thought from its logical justification? We would probably have no idea of the abstract number without prior experience of measurement, whether quantitative or geometric. For all that, is it legitimate to trace back the foundation of the number to the lived experience of counting and measurement? This would mean to confuse the historical genesis of thought with its logical genesis. This is also the reason why we should be wary of the role of intuition in mathematical philosophy: there is no doubt that no mathematics is possible without the mathematician's intuition, but this is not a logical necessity. Frege continues:

> The distinctions between a priori and a posteriori, the synthetic and the analytical, are not, in my opinion, concerned with the content of judgments, but rather with their justification. When someone says that a proposition is a posteriori or analytical, he is not making a judgement about the psychological, physiological or physical relationships that made it possible for the content of the proposition to be formed in consciousness; nor is he making a judgement about how someone else could have done it, perhaps erroneously, but rather about what, deep down, is the justification for holding the proposition to be true.
>
> Frege [50]

Psychology must therefore be discarded when it comes to mathematical truths: it is easier to understand why Frege systematically opposed seemingly innocuous practices such as those of Cantor, who sometimes appealed to the faculty of abstraction of thought: faculties such as the ability to abstract or generalize can lead us to apprehend the content of mathematical statements, but they do not guarantee their validity.

Beyond considerations of method, this epistemological and logical problem: *what is truth in mathematics and when should we consider a proposition to be true?* has a practical, even technical side. If Cantor, and before him the entire post-Aristotelian and Euclidean mathematical tradition was satisfied with incomplete, unfinished definitions that relied on evidence, it is also because he did not conceive that it was ever possible to go beyond a certain level of simplicity. Frege adheres to this point of view (not everything can be defined), but shifts its terms as he considers arithmetic truths to be a priori in the following sense:

> When a mathematical truth is at stake, the proof must be found and led back to the original truths *(Urwahrheiten)*. If, by this way, one arrives only at general logical laws and definitions, one has an analytical truth provided that one can exhibit the propositions on which the reliability of the definitions is based. When it is not possible to conduct the proof without using truths that are not of a general logical nature but which relate to a precise field of knowledge, then the proposition is synthetic.

And again:

> For a truth to be a posteriori, one asks that its proof cannot be conducted without reference to things of fact, that is to say, to undemonstrable truths, without generality, which include statements about particular objects. If, on the contrary, it is possible to conduct proof on the basis of general laws which are not susceptible and do not need to be proven, then the truth is a priori.
>
> Frege [50]

10.5 Arithmetic Statements

The general epistemological framework underlying the Fregean theses having been established, the exact treatment of arithmetic remains to be examined. Frege is familiar with the earlier literature. In the *Foundations of Arithmetic*, he takes a stand on the major theories of number that preceded him—an opportunity to point out their shortcomings and to justify his own approach. We will retain only a few of the most significant moments of his analyses.

Among the problems posed by numbers, there is a classical question that we have not yet tackled: that of the status and nature of arithmetic identities (Frege [50, Chap. 1 and 2]). This question plays a key role in Kant's work and gives Frege the opportunity to distinguish himself technically from Kantism. Let's consider the equality $2 + 3 = 5$. As elementary as it is, it poses delicate problems to the philosophy of mathematics. Essentially, the question is to know if an equality designates an identity between two expressions, two writings of the number 5 or the result of a process allowing us, a posteriori, to identify two a priori distinct contents.

In Kant's view, equality should not be conceived as an identity: it is the result of an act of synthesis that guarantees that two distinct contents of thought can be identified. The example of large numbers and equalities such as

$$135,664 + 37,863 = 173,527$$

shows that they do not result from an immediate identity of concepts and that the mediation of acts (of thought, of calculation) is necessary to justify equality.

For Frege, the Kantian thesis contradicts the need for reason to penetrate the foundations of science: in the end, these equalities refer less to the synthetic activity of thought than to the possibility of leading them back to deductive schemes that are part of a logical architecture of arithmetic statements. The Leibnizian point of view, although insufficient, seems preferable: for Leibniz,

> It is not an immediate truth that 2 and 2 make 4, assuming that 4 is 3 plus 1. It can be demonstrated as follows:
> Definitions:
>
> 1) 2 is 1 plus 1;
> 2) 3 is 2 plus 1;
> 3) 4 is 3 plus 1.
>
> Axioms: when the same is substituted for the same in an equality, equality subsists.
> Proof:
> $2 + 2 = 2 + 1 + 1$ (def. 1) $= 3 + 1$ (def. 2) $= 4$ (def. 3).
> Conclusion: $2 + 2 = 4$.

Frege validates this proof while noting that one should take into account the order of operations and implicit relations (those that the study of generalized numbers highlighted in the nineteenth century, such as associativity). One should write: $2 + 2 = 2 + (1 + 1)$; $(2 + 1) + 1 = 3 + 1 = 4$ and add $2 + (1 + 1) = (2 + 1) + 1$, a special case of the general axiom of associativity $a + (b + c) = (a + b) + c$.

Thus, by following Leibniz, 1 and the operation of increase by one (+1) make it possible to better grasp the analytical essence of number, while leaving in the shade the origin of these concepts of unit and increase (Frege [50, Chap. 3]).

10.6   Unity

Frege thus naturally turned his attention to the idea of unit, which he would study very closely. He notes the ambiguities that already exist in Euclid, where the word "μονάς" sometimes designates an object among those to be enumerated, a property of such an object or the number one. His criticisms become original when he examines the unit from the point of view of predicativity, which we have seen to separate concepts and objects according to him.

The most natural hypothesis concerning one seems to be to attribute to it the status of a concept, i.e. of a predicative function. I could affirm "Solon is one" as I affirm "Solon is wise", affirming Solon's being-one as his being-wise. But usage shows that this apparent predicativity of one is a dead end: if I am to give a meaning to "Solon is one", it is not a numerical meaning, but rather the meaning of "Solon is a whole". Moreover, the passage to the plural establishes the specificity of number statements: the plural of "Solon is one" and "Solon is wise" would be "Solon and Thales are two" or "Solon, Thales and Parmenides are three" and "Solon and Thales (and Parmenides) are wise". One is therefore not a concept in the logical, Fregean sense of the word. The word "one" will therefore be the proper name of an object whose nature remains to be clarified.

This technical elucidation constitutes, for unit and other numbers, the main object of the *Foundations of Arithmetic*. A preliminary step is to understand, not the numbers themselves, but the number statements. When I say, "There are four trees in front of me", I am dealing, in Fregean logic, with a statement about a concept (the one of "tree in front of me", which I can predicate from this oak, this pine, and these two poplars). A number statement is thus, generically, a statement about a concept. This very philosophical, Fregean terminology is sometimes difficult to follow today, and one must read the *Foundations of Arithmetic* step by step to appreciate its depth and understand how it emerges, slowly but naturally, from the analysis of arithmetic statements and the underlying thought processes.

Even if we have to anticipate its set-theoretic interpretation, which will be the subject of the next chapter, we can understand this conceptual status of number in the following way: a concept, for Frege, makes it possible to construct a multiplicity, a totality (that of the objects falling under this concept). A number statement assigns a cardinal number to this multiplicity. In the previous example, the concept of "tree in front of me" allows me to grasp as a distinct totality oak, pine and poplar; another concept ("poplars here", for example) would lead to another grasp. We will soon see how to attribute the number 4 to it, but it can already be seen that one of the interests of the Fregean approach is to put a strong emphasis on the empirical use of numbers in our ordinary judgment activities. Fregean logic in the *Foundations of Arithmetic* is a theory of the laws of pure thought, not a mathematical logic whose scope would be limited to the well-conceived statements of a formular language.

Chapter 11
Set Theory in Frege

The previous chapter, although it dealt with Frege's mathematical philosophy, its profound originality and its intrinsic difficulties, hardly solved the fundamental problem of establishing the status of numbers as objects in their own right.

Numbers would be objects, in the Fregean sense of the word, and number statements would be statements about concepts. Recall that in Fregean logic, concepts are essentially functions $f(\ )$, unsaturated entities that take a determination $f(x)$ only when the empty place, that of the argument, is filled by an object x. The *Foundations of Arithmetic* [50] seeks to construct propositional functions (unsaturated statements) allowing us to understand numbers as objects. Beyond its historical, mathematical, logical and philosophical importance, Frege's approach is exemplary of a certain type of mathematical creation where an intellectual project is at the origin of the production of ideas and concepts. The sources of the genesis of mathematical theories are indeed multiple (research programmes, concrete problems often motivated by the natural sciences...), but it sometimes happens, in some works of exceptional relevance, that one of the driving forces behind the most profound transformations of the corpus and the very idea of mathematics is the fruit of a reflection on the method itself and its justifications. Frege appears as such in the history of ideas alongside Descartes, Grassmann and, more recently, Grothendieck.

11.1 Mathematical Characterization of Number Statements

In any case, the problem is posed: "Once we have recognized that a number statement relates to a concept, we can seek to complete the Leibnizian definition of integers starting from 0 and 1". Frege proceeds in two stages, which can be described as mathematical and logical respectively: construct characteristic statements for the theory of natural numbers; acquire a logical understanding of their meaning, his

© The Editor(s) (if applicable) and The Author(s), under exclusive license to Springer Nature Switzerland AG 2020
F. Patras, *The Essence of Numbers*, Lecture Notes in Mathematics 2278, https://doi.org/10.1007/978-3-030-56700-2_11

ultimate aim remaining to base numbers and arithmetic on the pure laws of thought alone.

Frege's characterization of zero acquires for the first time its full modern conceptual extension: "To a concept is associated the number 0 if no object falls under the concept." Of course, Frege is not the first to have understood zero as, in contemporary language, the cardinal of the empty set, but his definition is part of a thought dynamic and a conceptual analysis very different from that which underpinned the philosophies of arithmetic from Aristotle to his contemporary Cantor. Echoing the Cartesian "I think therefore I am", the Fregean characterization of zero is based on an irreducible phenomenon, a borderline thought: that of a concept without object.

The two other moments of the Leibnizian definition of numbers are then the object of a similar treatment:

To a concept F is associated the number 1 if

1) It is not true in general that, whatever a is, a does not fall under F.
2) If of the two statements "*a* falls under F" and "*b* falls under F" we can conclude that a and b are identical.

The two axioms guarantee the existence and the uniqueness of the object falling under F. The passage from n to $n + 1$ is finally also explained in terms of the opposition concept/object: "To the concept F is associated the number $n + 1$ if there is an object a falling under F and if to the concept 'falling under F but different from a' the number n is associated."

11.2   The Logical Step of the *Foundations of Arithmetic*

The most difficult and most original step remains to be done: to acquire a logical understanding of what a sentence like "To the concept F is associated the number n" means. In agreement with most philosophers of mathematics, Frege notes that we do not and cannot have a representation or intuition of a number: we can of course have a representation of two objects, three letters, etc., but 2 or 3 and a fortiori large numbers are not objects that can be represented in themselves. In general, we do not have an intellectual intuition of mathematical objects that are always given to us through intermediaries. The case of small numbers and elementary geometry misleads us in this respect, making us confuse certain representations and immediate intuitions with their mathematical content.

Moreover, it is doubtful that we ever have an intuition of any theoretical concepts or objects whatsoever: what would be a full and authentic intuition of justice, of the notion of abstract group or of the big bang? But that is not the important thing: it is not a question of accessing a founding intuition of numbers, but of knowing how to recognize them in mathematical discourse and practice. Rather than an idealizing discourse, what science needs is a criterion for recognizing numbers, and it is by seeking to understand the springs of such a criterion that Frege will arrive at an authentically logical understanding of the foundations of arithmetic.

As for the criterion itself, it is already in Hume (1711–1776): "If two numbers can be so combined that one always has a unit that corresponds to each unit of the other, then we say they are equal." (Hume [63]) This is an informal version of the idea of one-to-one correspondence for which Frege refers, among his contemporaries, to Cantor, Schröder and Kossak [50]. He points out, however, that there are theoretical difficulties, which these authors underestimated and which he wants to confront. The notion of equality contains pitfalls that are too easily overlooked. Once again, Leibniz serves as a touchstone, who has sought to characterize these notions: "*Eadem sunt, quorum unum potest substitui alteri salva veritate.*" But should *eadem* be translated here as identical or equal? The two translations: "Two things are equal when they can be substituted one for the other while preserving the truth" and "Two things are identical when they can be substituted one for the other while preserving the truth" are statements that differ in their meaning. For example, $3+2$ is equal to 5, but there are statements where substitution is prohibited *salva veritate* ($5 \times 3 = 15$ versus $3 + 2 \times 3 = 15$, and an elementary school teacher will expect 5 as the end result of a calculation and not an unreduced expression like $3 + 2$). While these aporias can be fairly easily resolved, they are no less indicative of difficulties in principle.

11.3   The Copernican Arithmetic Revolution

Perhaps Frege's greatest intuition lies in this ultimate moment in the logical foundation of numbers, which consists in going against a naive idealism according to which the mathematical object exists from the outset and simply has to be recognized, identified, making instead of Hume's criterion of recognition the very logical origin of numbers. The philosophical scope of Frege's discovery cannot be underestimated, but its mathematical significance is at least as considerable. Frege was the first to grasp the nature of a mode of constitution of mathematical idealities that was to become an imposing figure in the twentieth century, without anyone after him having taken the full measure of this "Copernican revolution".

The example of geometry both sheds light on his approach and allows us to understand its universality. Two straight lines have the same direction if and only if they are parallel. Parallelism is therefore both a property of a pair of straight lines and a recognition criterion for the direction: if the departmental road that I take by bicycle to go to Lyon is parallel to the motorway that I would have taken if I had been driving, I know that I am going in the right direction. This criterion also makes it possible to give a definition: the direction of the straight line D is inseparable from the property "being parallel to D" and, for a straight line D', having the same direction as D is equivalent to belonging to the set of straight lines that satisfy this property. Of course, all this seems rather artificial and one can quite legitimately doubt having gained anything from this reformulation (X having the property P

being equivalent to X belonging to the set E of the Y's having the property P), but it has nevertheless transformed mathematical thinking from top to bottom by imposing, in a rather artificial way it must be said, the concept of set as the founding concept among all. In Fregean terms: D' has the same direction as D if it belongs to the extension of the concept "parallel to the line D". A concept extension thus makes it possible to define an abstract notion, by playing on the equivalence between criterion, concept and concept extension.

We can proceed in a similar way for numbers: equinumericity allows us to group concepts into classes. Recall that, in Frege's work, concepts have a collectivizing function. Each concept is associated with the totality of objects falling under the concept; the equinumericity relation therefore partitions the concept-sets into classes. Numbers are inseparable from these classes and, in the *Foundations of Arithmetic*, are identified with them: "The number associated to the concept F is the extension of the concept 'equinumeral to F'."

11.4   Frege's Ontology: 0 and 1

Unfortunately, as Russell observed, this beautiful construction is faulty and was not expected to survive long. Its underlying premise, that extensions of concepts are "objectivizing" (i.e. that the set of objects falling under a concept is an object on its own), stumbles over paradoxes. The best known are the impossibility of defining a "set of sets" or "the set of sets that are not elements of themselves", contradictory entities that illustrate the non-collectivizing character of the concept of set. Of course, these paradoxes may not seem to reach the body of mathematics, as they are based on "language games" rather than on the type of objects that mathematicians normally consider. These difficulties can be accommodated technically, but they partly ruin the Fregean program in its ambition to understand arithmetic in a purely logical way, based only on the laws of thought or language. To a certain extent, mathematics fixes here (as later with Gödel) limits on the pure play of concepts; this is in any case the lesson that Frege himself drew from the failure of his logicism to rigorously found arithmetic. Nevertheless, his work remains the foundation, largely unsurpassed, of the modern mathematical theorization of arithmetic.

The level of exigency of the Fregean thought, which led him to reject the logicist program as a whole because of paradoxes, is largely due to the almost philosophical nature of his project. For him, a definition is not an ontological argument: one cannot conclude from the absence of contradictions the actual existence of a mathematical object, it must still be produced to guarantee its legitimacy. Let us return to numbers: these are based on concepts and their extensions, but the certainty that concepts exist is a psychological certainty, not an analytical certainty. Also, to show that 0 and 1 exist according to the pure laws of thought, it is still necessary to produce purely logical concepts whose respective extensions of the classes of equinumericity are 0 and 1. We already know that it would be enough, to obtain 0, to consider a concept without object. Frege constructs it as an emanation of the laws of pure thought:

"As no object falls under the concept 'different from itself', I can declare: 0 is the number associated to the concept 'different from itself'."

Producing 1 may seem even more difficult: it requires a non-empty set, and one must therefore succeed in constructing, in a purely logical way, an object. The attempt would probably have seemed aporetic before Frege, since logic, the science of thought forms, is a priori incapable of producing content. But, "let's consider the concept—or, if we prefer, the predicate 'equal to zero'! Underneath it falls 0. Moreover, no object falls under the concept 'equal to zero but different from zero', so 0 is the number that returns to this concept". Thus is constructed, in a totally aprioric way, the successor of 0, the number 1, and this on the sole basis of the pure contradiction "different from oneself".

11.5   The Invention of the Empty Set Symbol

In addition to his various results and his research at the limits of pure forms of thought, one owes to Frege an elegant idea, whose origin is often unknown and philosophical meaning undervalued. Bourbaki, who has read Frege and cites his key works (the historical record of his *Set Theory* testifies a direct and precise knowledge of Fregean texts), makes an unquestionable judgment here: if the Fregean work is remarkable, "the symbols he adopts are not very suggestive, of an appalling typographical complexity and far removed from the practice of mathematicians" (Bourbaki [17]).

The judgment is not without foundation, for Frege's two-dimensional writing in the *Begriffschrift* [49] (to which Bourbaki's judgment refers) is indeed, in retrospect, clumsy, but, as we shall see, it does not go without a certain injustice. Indeed, it is to Bourbaki that we owe the popularization and standardization of a notation whose origin can be found first in Frege and consists in representing the empty set in the form of a crossed zero. There are some distinctions: in Frege, the symbol designates the class of sets isomorphic to the empty set, rather than the set itself, and the oblique line that crosses the 0 goes from upper left to bottom right, but this is a detail whose only meaning seems typographical. The principle of the notation is the same and is founded on the necessity to notionally distinguish between zero and the empty set or its isomorphism class, keeping however track of their close relationship. The symbol is present in Frege's *Grundgesetze* [52] as early as the table of contents and appears together with the logical construction of the empty set fairly quickly in the text. The analogy is striking, but never reported, since analyses generally stop at the statement by A. Weil (1906–1998):

> The part I had played [in the debates with Bourbaki] earned me the respect of my daughter Nicolette when I told her that I was personally responsible for the adoption of the symbol for the empty set, a symbol she had just learned to use at school. It belonged to the Norwegian alphabet, and I was the only one in Bourbaki who knew it.
>
> Weil [108]

All this would be anecdotal if the reference to the Norwegian alphabet, which has become the norm in historical records when it comes to reporting on the invention of the symbol, did not obscure one of the most attractive and conceptual features of Frege's use of the symbol. His use of crossed number symbols (*"durch einen schräg durchgehenden Strich ausgezeichnet"*) was thus not limited to zero, and the *Grundgesetze* introduced a similar symbol for the set-theoretical sibling of the counting number 1, in accordance with the logic of the Fregean set theory of seeing numbers as extensions of concepts. This accounts for the theoretical necessity to duplicate the number signs in the universe of sets, a necessity with far reaching epistemological consequences.

11.6   Hermann Weyl's Analysis and the Later Frege

We have already indicated in Chap. 8 how problematic Hermann Weyl continued to find the foundations of analysis until the 1930s. His book, *The Continuum* [109], contains, in addition to a set of ideas specifically related to real number theory, many critical reflections on post-Fregean set theory. The text obviously does not end the polemics and was largely outdated from the 1920s onwards, but it retains a historical, technical and above all, and this is what will retain our attention, epistemological interest.

Weyl's thought, whether mathematical or philosophical (or, very often, a mixture of the two), as original as it may be, is marked by the seal of classicism—where by classicism we mean less a certain conformism than the subordination of the means employed and the theses defended to the taste for measure and truth. *The Continuum* opens actually with the definition of truth: "A judgment affirms a state of affairs. If this state of affairs is realized, then the judgment is true; otherwise, it is untrue." The definition is anything but uncontroversial, for post-Fregean modernity would be tempted to define what is true in a purely formal, or even conventional way. Weyl thus seems to want to affirm from the outset his attachment to the meaning of mathematics, to its hold on reality.

He goes on with a crucial restriction, which distinguishes his approach from the logicist one: "A property is always affiliated with a definite category of objects, so that the proposition '*a* has that property' is meaningful, i.e., expresses a judgment and thereby affirms a state of affairs, only if *a* is an object of that category. For example, the property 'green' is affiliated with the category 'visible things'. So, the proposition that, say, an ethical value is green is neither true nor false but meaningless." (Weyl [109, p. 5]). Therefore, anyone who forgets that a proposition with a subject-copula-predicate structure may be meaningless is in danger of being confronted with absurdities, requiring an analysis of Russellian-type set-theoretic paradoxes.

His response to paradoxes is ultimately relatively simple and natural: to avoid the paradoxes of set theory, the semantic field of statements should be limited. For example, functions will be subject to have a domain of definition and, to avoid impredicative definitions and vicious circles, the use of quantization should be restricted (in the language of type theory, quantization is restricted to the first logical type).

Technical details are of little importance here, but it is worth noting the modalities of the Weylian opposition to Frege, which can be quite naturally interpreted as an opposition of style and taste between mathematics and logic. What makes the greatness, the genius, but also the mathematical limits of the Fregean approach lies in the will to bring arithmetic back to the pure laws of thought. Frege was too subtle, too profound, to identify this idea of pure laws of thought with a narrow conception of logic, but his attempt was nevertheless constrained by the will to avoid at all costs any recourse to mathematical objects and contents understood in the ordinary sense. This is the reason why there is no logical typification (no measurement of the level of abstraction in their definition) or categorization of objects (no delimitation of semantic/contentual fields) in his work: logic has always been the science of forms of thought, a science without contents. Fregean "objects" are therefore objects in a logical (or linguistic) sense: they are said to be objects solely for a certain symbolic use, which allows them to be substituted for empty places in propositions.

Weyl is too much of a mathematician to be satisfied with this approach, this reconduction of mathematics to logic, of "states of things" to "states of language". It is the whole meaning of the semantic restriction (in the sense of a restriction to meaningful utterances) that he proposes, and which makes it possible to eradicate set-theoretic paradoxes. It is for the same reasons that many mathematicians quickly feel uncomfortable reading the theses of the great actors of analytical philosophy, and this, in a way that is transversal to their different theoretical positions (Carnap, Wittgenstein, Quine,...). Beyond certain conclusions in analytical philosophy that are difficult to accept since they turn mathematics into a vast tautology, a science (or pseudo-science) without content, it is the whole project of reconducting thought to language that poses a problem.

This is indeed the observation to which Frege himself was led at the end of his career: a purely formal mathematical thought, without contents, seen through the sole prism of logic and linguistic structures is an aberration: "My efforts to clarify what is meant by numbers have failed. We are too easily misled by language and, in this particular case, the way we are misled is nothing short of disastrous." (Frege [53])

11.7 Weyl on Natural Numbers

The general idea that it is artificial to think of mathematics without any reference to some form of reality of its objects (through the notion of "states of things" in particular) applies quite naturally to numbers. Weyl's criticism of the approach that would bring the idea of number back to a system of axioms is emblematic of a series of criticisms addressed to the Hilbertian tradition such as those of Poincaré (1854–1912): "A set-theoretic treatment of the natural numbers such as that offered by

Dedekind may indeed contribute to the systematization of mathematics; but it must not be allowed to obscure the fact that our grasp of the basic concepts of set theory depends on a priori intuition of iteration and of the sequence of natural numbers." (Weyl [109, p. 24])

Weyl's analysis is quite modern in many respects and mixes ontological, logical and mathematical arguments. Axioms are supposed, in post-Hilbertian mathematics, to define the meaning of "it exists": ontology questions become logical ones. What exists is what the axioms imply. Fermat's last theorem for example, an open problem at Weyl's time, would thus hold if following from the axioms of arithmetic. Notice that this is contrary to intuition: when dealing with a statement like Fermat's theorem, asserting that a simple arithmetic identity can hold or not for some integers, we tend to spontaneously think that it has to be true or false independently of any axiomatic and logical framework. According to Weyl, the foundation of ontology on axiomatics actually has several drawbacks—for example, axioms have to carry a cognitive value to be relevant: mathematics is not a mere hypothetico-deductive game. It is ultimately untenable, even from a logical point of view. Weyl develops the example of real numbers. Following Dedekind, given two real numbers a, b, then $a < b$ if there is a rational number r such that $a < r < b$ and $b \leq a$ is true only if the axioms imply that there is no such rational r. However, it could happen that a system of axioms for arithmetic cannot decide the existence or the non-existence of such an r (consider for example the case where a and b are obtained as sums of very complex series only known to be convergent). In such a situation, for the axiomatic approach to ontology to be satisfactory, we would need to know that the axioms are consistent and complete, that is, that given a judgement U, one and only one among U and the negation of U is a logical consequence of the axioms. Even if such knowledge could be reached, one would need to appeal to the intuition of iteration in order to establish it. "But, from this intuition, we also directly obtain the fundamental arithmetic insights into the natural numbers on the basis of which the whole *mathesis pura* is logically constructed. And this undercuts the claim that the arithmetical axioms are mere stipulations." (Weyl [109, p. 18])

Let us leave aside for the moment the technical aspects and the problems of consistency and completeness—Weyl anticipates to some extent Gödel, whose proof of the incompleteness theorems relies entirely on the use of elementary arithmetic and the idea of recursion (or enumeration of the steps of reasoning) underlying logical proofs. Weyl's thesis (very close, it has been said, to Poincaré's) can be interpreted in terms of origins. Logic tends to spontaneously think of itself as a theory of origins: as the science of the rules of thought, it would be the condition of possibility of the other sciences. So be it, but it is counting for nothing the fact that thought pre-exists logic and that logic merely codifies existing discursive processes. It can, of course, be argued that these processes are universal and condition the very emergence of thought, but such a thesis is resolutely metaphysical and it is doubtful that logicians are ready to accept its most radical consequences (that is, a subordination of logic to the speculative theory of knowledge).

The example of numbers is enlightening: the very idea of demonstration implicitly supposes the notions of order and sequence (of the steps of a demonstration). It is, moreover, an infinitely more complex idea than that of number: it is doubtful that a child arriving at an age where he can abstractly understand numbers and their uses, including through complex properties such as the existence of operations, would be able to understand what a proof is. The axiomatic-formal approach and attempts to subordinate mathematics to logic have obscured what is obvious in such remarks.

Everything suggests that modernity and the future of mathematics will make it possible to better reflect on these questions of priority and on the complex relations between the genesis of concepts (their empirical and historical mode of emergence, partly conditioned by biological factors, partly by the possible existence of a normativity of rules of thought) and the existence of rigid forms of hypothetical-deductive reasoning codified by logic. The next chapters are partly devoted to this debate.

Chapter 12
Axioms and Formalisms

> *We believe in the reality of mathematics, but obviously when*
> *philosophers attack us with their paradoxes, we run and hide*
> *behind formalism and say 'Mathematics is just a combination*
> *of symbols deprived of meaning' [...]. Finally we are left in*
> *peace to go back to our mathematics and do as we have always*
> *done, work with something real.*
> *J. Dieudonné*
> *Quoted by J. Boniface [15]*

The Fregean work sought to free the theory of numbers from all roots in psychology and intuition to link it to logic alone and, more precisely, to a system of pure laws of thought. The paradoxes of set theory condemned this so-called logicist approach. This opened the way to the marked will, at the beginning of the twentieth century, to establish arithmetic and mathematics on other foundations or, at the very least, to try to elucidate what, in the Fregean project, had led to failure. The names of Hilbert and Gödel stand out among those attempts, which profoundly changed our understanding of the relationship between logic and mathematics, of the possibilities of formal systems and of the role of infinity in mathematics, while at the same time opening the way to new disciplines such as proof theory and making it possible to build a foundation on which computer science could then be born and develop.

With regard to set theory, to which we shall now return only anecdotally, "its destiny in the rest of the twentieth century was scarcely happier than the personal destiny of its creator, Georg Cantor."[1] Two misunderstandings are at the root of this situation, which are worth mentioning because they are linked to conceptual and technical difficulties in the treatment of arithmetic in the first half of the twentieth century.

"The first misunderstanding stems from the very success of set theory. What Zermelo grasped first was the possibility of using sets as the unique basis of the whole mathematical edifice [...]. This result was taken to be much more than what

---

[1]Dehornoy [38], whose analyses are followed hereafter.

F. Patras, *The Essence of Numbers*, Lecture Notes in Mathematics 2278,
https://doi.org/10.1007/978-3-030-56700-2_12

it is, namely a coding result, analogous for example to the possibility of coding the points of the plane by a complex number. Rough epigones have seen an ontological result where only coding is involved", thus giving set theory a disproportionate role which, for specialists in the field, "it does not claim to have" (Dehornoy [38]). In short, according to the contemporary conception, set theory is the delicate and as yet unfinished mathematical theory of infinity, from a Cantorian perspective far remote from the philosophical and foundational problems of the Fregean tradition.

The second misunderstanding lies in the meaning of Gödel's theorems (and other similar results), to which we will return later in the arithmetic framework, but which, for set theory, have the consequence that the classical presentation retained for example by Bourbaki in his *Elements of Mathematics* [16] is incomplete: "The ZF system of Zermelo–Fraenkel reflects an almost general consensus and its axioms express properties of sets which our intuition recommends to hold as true, but it is incomplete, lacunar; it does not exhaust the properties of sets" (Dehornoy [38]). This failure of an axiomatic system to account for the totality of the expected properties of a given domain of mathematical objects is undoubtedly the most important consequence, for mathematical philosophy, of the advances in mathematical logic in the twentieth century.

Frege's error, in his attempt to base arithmetic on the pure laws of thought embodied in an elementary theory of sets, is better understood today in the light of such observations; "[The early Frege] thought that logical induction[2] was 'categoric' (to put it in modern terms), i.e. that induction would exactly capture number theory, or that everything was said in Peano's arithmetic: this logical theory simply coincided, for him, with the structure and properties of numbers" (Longo [83]).

The modern understanding of the limits imposed on formal systems explains well, as we will see in this chapter, why it is structurally impossible to capture formally all the properties of numbers. However, it should be pointed out that Frege had already arrived at similar conclusions by following the path that was his: that of the pure laws of thought. The conclusions of the later Frege, who returned from logicism, ultimately brought him closer, albeit for very different reasons, to those of contemporary logicians who, following the example of G. Longo, strive to reevaluate for logical reasons the constitutive role of our intuition of space and time in mathematical thought (Frege [54]).[3]

Frege is also responsible for a very interesting discussion of the notion of axiomatic system.[4] In the classical conception, an axiom is "necessarily true" and refers to a field of scientific knowledge existing prior to axiomatization (the geometry of space, for example). The anteriority of the object domain to axiomatization guarantees the coherence of the system and makes the question of consistency superfluous: ontology guarantees logical validity. In a modern,

---

[2] On Peano's induction and arithmetic, see later in this chapter.

[3] See also Longo [83].

[4] G. Frege, "Logic in mathematics", in [54].

Hilbertian-type approach, the axiomatic system is intended to precede the object domain it defines—but then, as Frege points out, one slips from one conception of axiomatics to another without taking the trouble to distinguish conceptually between the two, which does not fail to interfere with all mathematical philosophy. An axiomatic in the Hilbertian sense (that of Euclidean geometry, for example), does not define its object domain, and only acquires an intuitive sense of geometric axiomatics through the choice of a model, a realization of the system of axioms in a given object domain (classical Euclidean geometry; the Cartesian product of copies of the field of real numbers...). In other words, a system of axioms is, like a function, structurally unsaturated: just as a function only takes on values when it is applied to a variable, a system of axioms only constitutes a field of mathematical objects when a model is given, i.e. when a realization is chosen.[5]

12.1 Dedekind

Dedekind occupies a very special place in the history of the foundations of arithmetic. More of a mathematician, no doubt, than Cantor or Frege, he made a decisive contribution to the emergence of the algebraic method characteristic of early twentieth-century German mathematics, which is known to have been the foundation of a large part of mathematical thought in the last century.[6] In the elementary theory of numbers, Dedekind is credited with a fundamental text, *Was sind uns was sollen die Zahlen?* [36], in which he approaches the problem of the constitution of numbers in a mathematically original way. Like Frege, his motivations are partly epistemological and logical:

> What is demonstrable must not be believed without proof. As obvious as this requirement may seem, it does not seem to me to be satisfied even with regard to the foundation of the simplest knowledge, namely that part of logic that deals with the theory of numbers, despite the most recent presentations that have been given. When I say that arithmetic (algebra, analysis) is only a part of logic, I express the conviction that I hold the concept of number to be totally independent of representations or intuitions of space or time: I rather hold it to be a direct emanation of the pure laws of thought [...]. Numbers are free creations of the human mind, they serve as a means to apprehend more easily and with more acuity the diversity of things.

---

[5]One can, of course, work in mathematics, as with functions that can be considered as objects in their own right, on systems of axioms "for themselves", in all their generality, but one is then dealing with objects of a higher level. Thus, working on a non-categorical system of axioms (that is, which does not determine its domain of objects) means working not on a system of mathematical objects (naive Euclidean geometry, for example), but on a system of theories (for example a set of geometries satisfying certain postulates, such as a set of geometries on archimedean fields). These questions are very well discussed from a logical point of view by Hilbert in his *Foundations of Geometry* [60].

[6]See Corry [32] and Patras [89].

When we follow precisely what we do by counting and enumerating things, "we are led to consider the faculty of the mind to relate things, to make one thing correspond to another, or to represent one thing by another: a faculty without which no thought would be possible".

The similarity of method and philosophical sensitivity with the Fregean approach stops there: Dedekind is a mathematician, and it is as a mathematician that he will try to construct the concept of number and to deduce its fundamental properties. The key idea of Dedekind, to which we will return in a more technical way in Chap. 15, is to reconduce numbers to a dynamic point of view, more ordinal than cardinal. Where Cantor or Frege understand number as based on equinumericity (the possibility of making the elements of two sets correspond bijectively if they have the same cardinality), Dedekind rather seeks to conceptualize the iteration process at work in actual enumeration. The key concept for him is therefore that of a chain: concretely, the data of a function f of a set to itself and successive images by f of an element of this set.[7] When the function is injective, Dedekind thus obtains a construction of the set of natural integers—a mathematical translation of the intuitive idea that all numbers can be obtained by iteration from the operation that associates to the integer n its successor $n + 1$.

Dedekind does not hesitate, and this attitude will be amply reproached by the partisans of the "purity" of the founding methods, to seek in the functioning of human thought the arguments that allow him to show that his theory has a content, that objects actually correspond to it. The existence of an infinite chain is ensured[8] by the faculty of our thinking to associate a concept with an object or a group of objects.

He further asserts this humanistic[9] conception of mathematics in a letter to Weber in 1888[10] and opposes the Fregean conception of number as a class of sets:

> By the cardinal number *(Anzahl, Cardinalzahl)*, one must understand not the class (the system of all equivalent finite systems), but rather something new (corresponding to this class) that the mind creates. We are of divine essence and undoubtedly possess the power to create, not only in material things (railways, telegraphs), but especially in intellectual things.

---

[7]We refer to Dedekind's original text, or to one of its detailed comments, for a methodical description (which here would be of limited interest for our purposes). See for example Boniface [15].

[8]We have already had occasion to mention this in connection with the third man argument, taken up on this occasion by Dedekind.

[9]In the sense that human thought plays a central, demiurgic role.

[10]Quoted in Boniface [15, p. 24].

12.2 Peano

Dedekind's approach, although non-formal in the modern sense since it relies on elements of cognitive psychology to justify the existence of object domains (object systems, functions, chains...) underlying arithmetic, leads him, by its resolutely mathematical character, to isolate fundamental principles on which to build arithmetic. In other words, the keys to an axiomatization of arithmetic were already in Dedekind's hands in 1888, but it was not until Peano,[11] 1 year later, that this axiomatization was enunciated as such.

Faithful to the principle retained for this work of limiting references to technique to the strictly necessary, we will only give an informal presentation of Peano's system of arithmetic[12] in the form of a quotation of Couturat (1868–1914) [33]:

> Mr. Peano admits as primitive, indefinable ideas zero, the idea of number (positive or null integer), the idea of the number following another. He then poses the following principles or axioms:
>
> 1: 0 is a number
> 2: Any number is followed by a number
> 3: Two numbers followed by the same number are equal
> 4: 0 does not follow any number
> 5: Principle of complete induction: if a proposition is true for the number 0 and if, being true for the number n it is also true for the next one, it is true for all positive or null integers.
>
> We can replace in these axioms 0 by 1; we then define the idea of a non-zero positive integer [...]. Mr. Peano's theory differs from that of Mr. Dedekind only in that it admits as an axiom the principle of complete induction, which the German scientist demonstrates starting from the notion of chain, and in that he does not claim to define the positive or null integer, but to enumerate its characteristic properties.

In spite of the reference to Dedekind for the authorship of the system of axioms, it is worth paying tribute to the work of Peano whose methodological clairvoyance, clean execution, a certain notational and conceptual elegance, and the adequacy of means and ends make him an indispensable reference in the history of the construction of modern mathematical thought. It is no coincidence that history has most often retained the system of standard axioms for arithmetic as Peano's Arithmetic.

---

[11] Who readily acknowledges the influence of Dedekind on his work.

[12] On Peano's axioms approached through the filter of the history and philosophy of mathematics, see for example Potter [99, p. 82] or Godefroy [56, p. 227].

12.3 Hilbert

Peano's axiomatization of arithmetic opened up a field of reflection that developed until Gödel's work and beyond, as evidenced by contemporary research on large cardinals and infinity. The greatest historical figure of the axiomatic movement and the concomitant reflection on its meaning and its technical and philosophical scope is undoubtedly Hilbert.

The paradoxes in set theory and the failure of the accompanying Fregean logicism had profoundly shaken the certainties of the mathematical community with regard to the problem of foundations. In a whole part of the community, the set-theoretical approach and the nonchalance with which actual infinity is accepted there provoked uneasiness and mistrust, leading, for example, to the construction of Brouwerian intuitionism, of which H. Weyl was one of the exponents. Arithmetic, because of its architectural role for all of mathematics, but also because it has always been at the heart of the relationship between the finite and the infinite, has naturally been at the centre of these debates, both mathematical and philosophical. Thus Weyl asserted in *The Continuum* his firm conviction that the idea of iteration is an ultimate foundation of mathematical thought (Weyl [109, p. 48]). In agreement with Poincaré he observes that, while it is true that the fundamental concepts of set theory can only be grasped through the intuitions of iteration and natural numbers, it is useless and misleading to revolve around them and try to provide a foundation through set theory for the natural number concept.

In concrete terms, Weyl rejects the foundational approach through set theory, instead preferring to build mathematics on a few fundamental intuitions and a set of processes. Whether one agrees with him or not, the questions he raises are essential and relate to the key problem of origins: on which fundamental processes should the mathematical edifice be built?

This type of reflection was far from being isolated. They took place in a variety of ways depending on the viewpoint and the mathematical and philosophical interests of the participants. The early Couturat[13] [33] also opposed Dedekind for largely philosophical reasons:

> That Dedekind succeeded in defining the whole numbers logically, that is to say by means of pure concepts and abstract relations, without any recourse to intuition, is what we allow ourselves to doubt [...]. The question is worthy of careful consideration, as there are few questions whose solutions are more important for the philosophy of mathematics and for the theory of knowledge in general [...]. As we shall see, it is in the idea of number itself that we believe we find the intuitive and synthetic element that makes arithmetic judgments refractory to pure logic and irreducible to analytical judgments.
>
> Number is not a concept, but an intuition; in other words, one does not define a number, one can only show it. This thesis, which could be justified by many other reasons, is in line with Kant's doctrine, according to which mathematical truths, even those of pure arithmetic, are a priori synthetic judgments. It is perhaps not without interest to note that the tendency

---

[13] He was then to become an advocate of symbolic logic in France.

of modern mathematicians to reduce the primitive data of their science to purely logical notions [...] only verifies and consolidates Kantian theory.

Hilbert, also strongly influenced by Kantism,[14] although violently opposed to intuitionism, was to be confronted with the same difficulties and sought to answer them once and for all at the end of a real bet: to try to base all the mathematics of his time on the finite, without accepting the sacrifices to which intuitionism leads. The latter was indeed reluctant to use many fundamental methods (excluded third, axiom of choice...) commonly used in the treatment of the continuum and more generally in all analysis.

Hilbert's reflection on the relationship between arithmetic, logic, the axiomatic method and fundamental problems starts at the beginning of the twentieth century and follows his axiomatization of geometry (1899) and the idea that the non-contradiction of geometry finally reduces to that of the theory of real numbers. The 1905 article [61] marks a turning point: Hilbert presents a first attempt to prove the non-contradiction of arithmetic. Hilbert's ultimate ambition was to guarantee the validity of the mathematical edifice; it took an astonishing foresight to first conceive the possibility of trying to demonstrate that a mathematical theory can be free of contradictions, and then to conceive the idea of a proof strategy. These methodological, strategic and conceptual advances were going to revolutionize our conception of mathematics by bringing to the fore scientific and philosophical research themes that had hitherto been unexploited or even unsuspected, for example as regards the nature and mathematical structure of proofs, or the distinction between mathematical language (the language in which mathematics is written) and metalanguage (the language that allows us to talk about mathematics, its structure, its semantics, etc.). The essence of the Hilbertian method is perhaps already in the conclusion of this 1905 article:

> I think that we can give a rigorous and entirely satisfactory foundation to the notion of number, and this by a method that I will call axiomatic and whose fundamental idea I want to briefly develop in what follows. Arithmetic is often seen as part of logic, and traditional fundamental logical concepts are generally presupposed when establishing foundations for arithmetic. However, if we look carefully, we realize that in the traditional exposition of the laws of logic some fundamental arithmetic notions are already used: for example, the notion of set and, to some extent, also the notion of number. We are therefore caught in a circle, and that is why a simultaneous development of the laws of logic and arithmetic is required if we want to avoid paradoxes.

The idea is therefore twofold: to develop in parallel logic and mathematics by renouncing the logicist program in the strict sense; to use as the foundations of arithmetic those fundamental notions that predate or coexist with primitive logical notions. At the forefront of these, Hilbert places then the notions of object of thought (*Gedankending*), object (*Ding*) and the idea of combinations (*Kombinationen*). He later retained the conviction that the essence of mathematical thinking is based

---

[14] According to modalities different from those found in Couturat and analyzed by J. Boniface [15, sections 6.II.3 and 6.II.4].

on the intuition of the discrete and its combinations. Hilbertian "Kantism" in mathematics can be understood as a replacement of Kant's intuition of space and time at the foundation of mathematics by the intuition of the finite.

> A serious study leads us to recognize the existence, in addition to experience and thought, of a third source of knowledge. Even if today we can no longer agree with Kant in detail, the most general thrust of Kant's theory retains its scope: to define the a priori attitude and thus to study the condition of possibility of knowledge. In my opinion, this is, for the most part, what my work on the principles of mathematics has achieved. The a priori resides there no more and no less than in a fundamental attitude that I would willingly characterize as finitist: we are already given in advance with precision something in the representation, certain extra-logical concrete objects that, intuitively, are, as an immediate experience, present before any thought.[15]

The result is an original and atypical mathematical philosophy, not entirely formalist or logical because of this residue of Kantism. Hilbertian mathematical ontology, in particular, does not limit itself to conceiving the existence of mathematical objects in the light of the non-contradiction of formal systems, as is often believed and said, since this existence is based on an irreducible intuitive foundation, that of the intuition of objects, symbols and their combinations.[16]

This frequent idea that Hilbert's ontology is limited to non-contradiction must be further nuanced by another factor, partly technical, which is theorized and present in Hilbert's discourse from the *Foundations of Geometry* onwards: Hilbert realized very early that a system of axioms is rarely categorical (it rarely determines a domain of objects univocally). Axiomatization (e.g. of geometry) therefore usually only captures some, but not all, properties of a mathematical domain. The corresponding ontology is therefore relative: the triangle of Euclidean geometries is not the triangle of general Riemannian geometries. This is all the more easily accepted as our intuition of these two families of triangles is quite different, but things can be more subtle and complex. Without going very far in complexity, the arithmetic number is thus not the same according to whether or not we allow the principle of induction, since Peano's arithmetic without the principle of induction lends itself easily to the construction of generalized numbers.[17] The ensuing ontological relativism is not without problems because, if there are ultimately as many families of numbers as there are models of arithmetic, in which theory will we recognize the "true" numbers, assuming that such a question still makes sense?

Beyond these philosophically delicate questions, Hilbert's program, proving the consistency of arithmetic by finite processes, for example explicit calculations on integers, immediately comes up against a problem of principle related to the distinction between mathematics and metamathematics. Admitting finite procedures, as Poincaré will point out, means admitting integers and, therefore, what sense can

---

[15]Hilbert [62, pp. 485–86] quoted by J. Boniface [15].

[16]It is nevertheless a minimal ontology, far removed from ordinary mathematical ontology. This poverty of Hilbertian and post-Hilbertian ontology is well analyzed in the work of G. Longo, to which we refer.

[17]See e.g. Godefroy [56].

a proof of the consistency of arithmetic have, since it admits what it wants to prove, the legitimacy of the use of numbers and the associated reasoning and calculations? In fact, the Hilbertian program admits the validity of elementary calculations with numbers[18] and aims to extend the field of validity of these calculations to quantized arithmetic (for any integer $n...$, there exists an integer n such that...), but this distinction between necessarily valid proto-arithmetic that predates mathematicians' arithmetic and mathematicians' arithmetic, the consistency of which would have to be demonstrated, has something partly artificial and philosophically problematic in Hilbert's time. Gödel's theorems, which will now be discussed, have largely elucidated such difficulties.

12.4 Gödel

The fundamental idea underlying Gödel's theorems, although technically delicate to exploit, is simple and is in line with the Hilbertian project to bring all mathematics back to the finite and to natural numbers in their most elementary use.[19] Arithmetic and, more generally, mathematics, are coded symbolically. Thus, any statement, any arithmetic proof is a series of symbols that can be coded in the arithmetic itself.[20] The remark applies to any formalized mathematical theory, i.e. defined by a finite set of symbols and formal rules like Peano's arithmetic;[21] such a theory is codable in arithmetic and the basic problems (non-contradiction, completeness...) thus appear in the end as arithmetic problems![22] In other words, metamathematics can, after Gödel, be understood as a part of mathematics, albeit of a rather specific type and whose interests are only marginally similar to those of most classical branches of mathematics.

In any case, since Gödel, it is possible to encode in arithmetic the metatheoretical notion of demonstrability. This technique of arithmetization of the properties of demonstrability of statements leads directly to Gödel's theorems: a formalized theory of numbers, assuming that it is non-contradictory, is incomplete, there are always undecidable statements, i.e. not demonstrable and whose negation cannot be demonstrated either. An arithmetic content can be given to them, as well as

---

[18] At least in a rudimentary form, that of symbol manipulation.

[19] See Nagel and Newman's book [87] for an elementary and pleasant introduction to Gödel's work.

[20] One method is to use the uniqueness of the decomposition of a natural number as a product of primes. In practice, the general idea is to code a sequence of symbols such as $1 + 5 = 6$ by an integer in a univocal way, so that the knowledge of the integer allows us to go back to the sequence of symbols. Thus, by associating once and for all distinct integers $n_1, n_2, n_3, n_4, n_5, n_6$ to the symbols $1, +, 5, =, 6, -$ and using the sequence of prime numbers $2, 3, 5, 7, 11, 13...$, the identity $1 + 5 = 6$ will be coded by $2^{n_1} \times 3^{n_2} \times 5^{n_3} \times 7^{n_4} \times 11^{n_5}$, while $6 - 5 = 1$ will be coded by $2^{n_5} \times 3^{n_6} \times 5^{n_3} \times 7^{n_4} \times 11^{n_1}$. See Godefroy [56].

[21] Theories involving non-finite rules fall outside the scope of Gödel's work.

[22] In a very specific sense, as we shall see later.

an intuitive content: they can be chosen to be "true" in a non-formal sense (their interpretation is true in the universe of "intuitive integers", those with which we are familiar in our non-formalized use of numbers) (Godefroy [56, p. 162]). Better still: the formal coherence of arithmetic is not demonstrable by arithmetic procedures alone, which ruins in passing Hilbert's hopes of a finite deduction of the coherence of existing mathematics: "This marks the death of the ultimate foundation of mathematics on the absence of meaning, on a potentially automatable calculation of signs." (Longo [85])

Intuitively, the conclusion that emerges from Gödel's results and their method of proof, as well as from the subsequent work, is quite simple and natural. To prove advanced metatheoretical results (coherence, completeness...) about a given mathematical theory, one needs more powerful tools of deduction than those that the theory itself allows. This result will not surprise any mathematician. It is for example one thing to solve an equation or a system of equations, another thing to elaborate the "metatheory" governing the obtaining of solutions. The history of mathematics abounds with examples in this sense: the solution of linear systems of equations with scalar coefficients, real or complex, requires the passage to matrix algebras, which are noncommutative; Galois' theory, which predicts the unsolvability of polynomial equations by radicals, is of a theoretical and conceptual complexity that is out of all proportion to the naive theory of solutions of polynomial equations... It is indeed this same heuristic that manifests itself in mathematical logic; the "theory of the theory" is structurally more complex than the theory itself.

Another more radical conclusion emerges: "The (formal) principles of proof do not have the expressiveness of the principles of construction (order and symmetry) that have produced the conceptual structures of mathematics." (Longo [85, p. 35]) In Kantian language: the historical construction of mathematics proceeds largely, both at the level of objects and methods of proof, by synthesis. The diversity of materials, intuitions, representations, objects of mathematical thought is then subsumed under generic, fundamental concepts. Axiomatization is one of the modalities of this process, but nothing in actual mathematical thinking is free of synthetic moments. Gödel's theorems also condemn the formalist ambition to reduce mathematics once and for all to a purely analytical science, and if the ambition can subsist it is now rather as the infinite task of reducing synthetic moments of knowledge to analytical moments, in a way that today allows parts of the classical mathematical corpus to be encoded in computer programs.

12.5   Moderate Platonism

The philosophical repercussions of Gödel's work are many and important, but Gödel himself was responsible for trying to understand some of them in a speculative way. His reflections, largely supported by the example of arithmetic, led to a relegitimization of Platonism in mathematical philosophy[23]—Platonism that had been challenged and decried both by the proponents of a logical approach to mathematics, from Russell to analytical philosophy via Carnap, and by the structuralist current that had wanted to radically free itself from any reference to mathematical objects conceived as idealities.

Alain Connes has given a fairly faithful account of the guiding ideas of Gödelian Platonism.[24] According to him, formalism only reflects one aspect of mathematics whose limits are shown very clearly by Gödel's theorems. "This convinced him, because of the necessary distinction between truth and provability, for example, to separate raw mathematical reality [. . .] from the axiomatic and logic-deductive system that we develop to understand it." (Connes [31, p. 37]) True to Platonism, he maintains that mathematics has an object, just as real as that of the sciences, but not localized either in space or in time. Gödel's works showing that we have to distinguish between arithmetic truths and the logical consequences of axioms hint therefore at the existence of an archaic reality that escapes the mode of exploration given by the axiomatic and logic-deductive method.

Thus, Connes takes up the major theses of Platonism in its most common form within the mathematical community, but contains (in addition to the specific quality of the testimony of one of the most profound and creative mathematicians of the recent period) an original ingredient compared to the traditional expositions, which is the reference to Gödel's theorems and arithmetic statements.

"Archaic realism", in the sense that there is, in arithmetic, an archaic, primordial reality, pre-existing to mathematical theorizations is a position generally rather qualified as "moderate (Gödelian) Platonism", since the principle is already found in Gödel. The adjective "moderate" refers to an ontologically weak version of Platonism where all mathematical objects do not necessarily have the same reality (the status to be accorded to actual infinity remains for example largely problematic in a post-Gödelian perspective). The underlying idea lies in the ontological robustness of "intuitive" numbers and in the impossibility of capturing their properties by purely formal methods. Since logic had to give up, after Gödel, defining and accounting for them without implicitly calling on them and on stronger properties than those it would like to account for, it is that, in one way or another, numbers "exist" prior to these efforts.

The arithmetic statements characteristic of post-Gödelian mathematical logic (relating for example to diophantine equations: polynomial equations with integer coefficients and multivariables, of the type of Fermat's equations),[25] testify well to the significance of this renewal of Platonism on technical grounds. It can be demonstrated (with tools more powerful than those of Peano's arithmetic, but whose

---

[23] See Patras [92].

[24] Given the dialogical form of the text, A. Connes does not give explicit references to support his theses, but their content adheres quite precisely to what is generally considered the fundamental thesis of Gödelian Platonism.

[25] See for example Godefroy [56, Chap. IX].

validity is guaranteed to us by mathematical intuition) that some of these equations admit no solutions, but that this inexistence of solutions cannot be demonstrated in Peano's arithmetic. This is actually an illustration of the incompleteness theorems: the exhibition of a true result for "intuitive integers" that is undecidable in Peano's arithmetic.

From a strictly logical point of view, one could try to argue that, since there are properties of these equations that are not demonstrable in Peano's arithmetic, they are neither true nor false for his concept of number, and that more generally positive integers only ever exist in reference to a formal system that defines them. However, this is precisely a position that Platonism will reject: since these equations are demonstrably unsolvable, it is because formalism fails to capture the true nature of numbers. It is to turn things upside down to declare that intuitive numbers would have a less real existence than those of formal theories! As delicate as these debates are, this is in any case the meaning (very interesting and deep philosophically) of the theses of moderate Platonism.

12.6   Transfinite Numbers

The misunderstandings that hamper the understanding of arithmetic, whether intuitive or formalized, are due to its immediacy and obviousness. In our daily practice of numbers, we know that the equations of arithmetic (without quantifiers) are essentially decidable. It is "enough" to recursively try all the numbers in an arithmetic equation to finally find the solution, if it exists. Of course, this "it is enough" poses considerable difficulties. In particular, one cannot be sure that no solution exists if none was found after a finite number of steps of the algorithm...

It is precisely the actual infinity, that is, the necessity, in mathematical thought, to consider all the natural numbers simultaneously (for example in a universal quantification: "for any positive integer n, the property $P(n)$ is true") that leads to the aporias encountered by Frege and Hilbert. Mathematicians have long thought that the principle of induction would be sufficient to capture the properties of infinity necessary for the rigorous mathematical construction of arithmetic. It is indeed one of the lessons of (the failure of) the Hilbertian program, of Gödel's theorems and of later works (Gödel and Cohen, in particular), that this intuition is false. Just as there is, in Cantor, a true zoology of the infinite, the principle of induction admits levels of which the classical induction on the set of natural numbers is only the first, insufficient for example to demonstrate the coherence of arithmetic.

The idea of the transfinite refers to the Cantorian arithmetic of the infinite; as soon as one accepts the actual infinite, one can "count", calculate and demonstrate by recurrence beyond the finite. In the case of classical induction, the underlying key mathematical property is that any non-empty subset of the set of natural numbers has a smallest element. However, this property is not specific to integers and allows the domain of numbers to be extended consistently. Indeed, if one asks that this principle be retained, the sequence of positive or null integers extends naturally. The first transfinite number (beyond the usual integers) is denoted by ω, it is followed by $\omega + 1, \omega + 2, \ldots, \omega + \omega, \ldots, \omega^2$ etc.,[26] where the notation $\omega + 1$ designates for example the smallest transfinite number beyond the integers and ω.

The use of transfinite numbers—this is one of the lessons of the post-Gödelian inheritance in mathematical logic and modern set theory—makes it possible to demonstrate arithmetic properties which remain inaccessible to Peano's arithmetic, transfinite induction (i.e. on transfinite numbers) being irreducible to the usual induction. The convergence of Goodstein sequences gives a good example of this.[27] The latter is an easily described arithmetic property which we immediately recognize as stating a property of "intuitive" numbers (any debate on the relative character of the idea of number to an axiomatic universe would appear here very artificial), and yet the proof of the theorem escapes, as can be demonstrated, Peano's arithmetic!

We note in passing that such examples have an epistemological meaning that goes beyond Gödel's theorems in their original presentation. In the latter, the "true" utterances that could not be demonstrated still had a structure quite close to that of the liar's paradox, a structure of logical type therefore, whose concrete meaning for arithmetic could leave us doubtful. Transfinite induction even makes it possible to demonstrate the non-contradiction of arithmetic![28] These different results and observations relativize the scope of Gödel's theorems in their most common meaning or, more precisely, give them a truly mathematical dimension and content.

As for the impact of these ideas from logic and set theory on mathematics as a whole, the observation made by specialists in the field is rather reserved. Thus, in an article on the impact of the incompleteness theorems, the logician and philosopher S. Feferman concludes that, "I have conjectured and verified to a considerable extent that all of current scientifically applicable mathematics can be formalized in a system that is proof-theoretically no stronger than PA [Peano Arithmetic]" [47].

Feferman also distinguishes between two inexhaustibilities in mathematics: besides the one unraveled by Gödel's incompleteness theorems, there is a potential infinity of propositions that can be demonstrated in a given axiomatic S, but "at any

---

[26]Our presentation of Goodstein's transfinite numbers and sequences is taken from Dehornoy [38]. On this subject, see also Godefroy [56].

[27]Their definition is based on the iterated expansion of positive integers in base p. In base 2, for example: $26 = 2^{2^2} + 2^{2+1} + 2$. For $q \geq p \geq 2$ we define, for any integer $n \geq 0$, $T(p, q, n)$ as the integer obtained by replacing everywhere p by q in the base p expansion of n. For each integer $d \geq 0$, we then define the Goodstein sequence of base d as the sequence of integers $g(2), g(3), \ldots$ satisfying $g(2) = d$ then, inductively, $g(p + 1) = T(p, p + 1, g(p)) - 1$ if $g(p)$ is non-zero and $g(p + 1) = 0$ otherwise. Starting from $g(2) = 26$, we find $T(2, 3, 26) = 3^{3^3} + 3^{3+1} + 3 = 7625597485071$. It seems clear that the sequence thus obtained tends towards infinity extremely quickly, and yet Goodstein has shown that it always tends towards 0! See Dehornoy [38], to whom we also refer for a proof.

[28]Gentzen, 1936, see Godefroy [56, p. 182]. Gentzen's result does not contradict Gödel's theorems since transfinite induction is not part of arithmetic in its usual axiomatization.

moment, only a finite number of them have been established. Experience shows that significant progress at each such point depends to an enormous extent on creative ingenuity in the exploitation of accepted principles [in S] rather than essentially new principles. But Gödel's theorems will always call us to try to find out what lies beyond them."

Chapter 13
The Brain and Cognitive Processes

Our mental activity originates in the brain and is inseparable from its physico-chemical processes. So be it. However, this observation leaves open considerable problems, both for the empirical sciences (biology, chemistry, physics) and, from the more abstract point of view, for the relations between thought and its material roots.

The study of numbers lends itself well to this type of analysis, for various reasons which will be discussed in this chapter. To a large extent, the key problem is to understand the constraints that our neural architecture imposes on mathematical activity. Paradoxically, the richness of the possible meanings and uses of numbers, the fact that they are part of very primitive or even animal activities as well as some of the most sophisticated levels of abstraction (in contemporary arithmetic), makes their study delicate from the point of view of cognitive analysis. On the other hand, this polysemy of numbers, combined with a certain form of conceptual unity, is interesting in that it highlights the way in which very different cognitive processes can collaborate in the creation of idealities: a phenomenon that clearly shows the impossibility of reducing mathematics to a single norm, be it of a logical, philosophical or psychological nature.

13.1 The Real Distinction Between Soul and Body

Descartes, to whom is due the thesis of a real distinction between soul and body, is often misunderstood and judged by the yardstick of later problems and points of view in which he would have hardly recognized himself. This thesis, as in his *Treatise on the Passions of the Soul*, which studies, with the resources of the time, the problem of the problematic union of soul and body, is undoubtedly not for nothing the work of one of the best scientific and mathematical minds in the history of mankind.

© The Editor(s) (if applicable) and The Author(s), under exclusive license
to Springer Nature Switzerland AG 2020
F. Patras, *The Essence of Numbers*, Lecture Notes in Mathematics 2278,
https://doi.org/10.1007/978-3-030-56700-2_13

What is it all about? The inhomogeneity of thought to matter. Mathematical ideas (Cartesian geometry or Fermat's theorem, for example) are irreducibly immaterial, as is thought more generally. Of course, understanding these ideas requires brain activity, but the reductionist view that the content of Fermat's theorem could be reduced to a few synaptic connections poses considerable problems and ultimately impoverishes the theory of knowledge.

Several facts militate against this reductionism: the collective dimension of knowledge; the possibility of reactivating extinct knowledge through the reading of forgotten manuscripts, or the impossibility of having an intuitive access to entire areas of mathematics other than through symbolic activities whose meaning is not directly played out in the cerebral activity of apprehending symbols, but in the relations between these symbols and a collective and historically constructed universe of contents. In a more prosaic way, physico-chemical reductionism would also condemn science to no longer be rigorous, as our current understanding of neuronal activity clearly shows that randomness plays a determining and irreducible role in it.

Accepting the thesis of a real distinction between thought and matter, however, by no means closes the debate: their problematic union and the need for thought to be anchored in a material support open up many difficult questions which cognitive neuropsychology is now tackling with the support of modern technology, but which knowledge theory has always been confronted with, albeit in other ways.

13.2   On Faculties, Time and Space

As far as numbers are concerned, a classical thesis, found particularly in Kant, states that numbering is based on our apprehension of time and its flow, which spontaneously creates in us the ideas of order and succession. The constitution of idealities would thus be based on our cognitive faculties, and the horizon of the latter would also be the ultimate horizon of the former: the domain of ideal objects cannot exceed that of our faculties.

More radically, our theoretical knowledge would be structured by these faculties, of which it would constitute, in the Kantian language, an a priori synthesis. Kantism has been much criticized for its limited conception of mathematics as an a priori science of space and time. In the case of numbers and arithmetic, it can be criticized above all for having established as a norm one point of view among others, that of duration and succession.

Modern mathematics, infinitely more abstract than that of the eighteenth century, leaves little room for a naive conception of the relationship between ideal objects and faculties. The fact remains that day-to-day mathematical work, especially in its creative phase, is largely based on the ability to give flesh and body to ideas in the form of representations of different levels (geometric objects, numerical symbols, axiom diagrams, etc.). Bats, bears or sharks, supposing them to be equipped with a brain capable of abstraction but retaining their own organs and sensory abilities, would probably have a different geometry from ours.

In a very concrete way, this anchoring of creation and mathematical work in representative and intuitive faculties seems to largely explain the existence of "styles" and tastes in mathematics: each person chooses, when he or she decides to make a career in mathematics, the type of objects and theories that best correspond to his or her abilities. In the field of discrete mathematics alone, which covers arithmetic and combinatorics, there are thus clearly differentiated scientific attitudes and very different approaches to mathematics.

13.3 Learning Arithmetic

The different mathematical and philosophical conceptions of number have, in fact, their cognitive counterpart. The child who learns to distinguish the very first numbers $(1, 2, 3 \ldots)$, who learns to count on his fingers, who links this learning to succession, to iteration, who learns to measure distances, makes complex syntheses in a largely unconscious way. In terms of cognitive neuropsychology, each of these learning processes seems to take place in distinct areas of the cerebral cortex: visually perceiving small quantities is a distinct process from verbalization, methodical counting, and the perception of measured time or space.

There must therefore be a spontaneous synthetic activity that allows the child to understand, or rather to gradually get the intuition, that the same regulatory concept (number) is at work in these multiple processes. Of course, these phenomena are not specific to numbers: most of our activities, both practical and theoretical, especially when they are verbalized, brought to language and referred to ideas and concepts, would testify, if they were precisely analyzed, to this same synthetic and associative faculty of thought.

In mathematical work, these phenomena manifest themselves on a daily basis but often go unnoticed because they are not thematized as such. The first moment of the work is to familiarize oneself with mathematical objects, to represent them, to be able to better apprehend them. However, the work only becomes truly fruitful when the imagination begins to create links between them. The process is slow and progressive: the mind first guesses similarities, analogies, for example between the behaviour of such and such a solution of a differential equation and such and such a probabilistic object; in the solution of a problem of analysis and geometry. This similarity, when it is founded in reason, then crystallizes in the form of a common and transversal, unifying notion, in the way in which, in the child, the counting number and the measuring number are going to be found in the common figure of the abstract, general number.

At the psycho-cognitive level, which interests us here, these processes are quite mysterious. While it is fairly easy to admit that regions of the brain are specialized in perception, language, symbolic activity…(Dehaene [37]), the process of synthesis giving all these activities the unity they acquire in thought and sometimes in the purity of concepts is difficult to grasp. The previous chapters of this book have moreover emphasized the difficulty for mathematicians and philosophers to agree on the definition and origin of numbers: proof that the synthesis carried out spontaneously by the child is far from being evident! The extraordinary flexibility of the brain and thought, which allows such synthetic acts on the sole basis of analogies, is very probably one of the most fertile and profound sources of discovery and progress of knowledge.

13.4   The Organ of Numbers

The preceding analyses may lead us to believe that, in the end, numeration, however complex the paths of its mental constitution may be, proceeds mainly from channels and origins in agreement with the various classical theories of numbers (ordinality, cardinality, anchoring in temporality. . .). Cognitive psychology has, however, long ago brought to light the existence of much more primitive processes which are to a certain extent transversal to these theorizations—transversal in the sense that perception, in its immediacy, is always below or beyond the theoretical schemes that can be superimposed on it. Thus, following Dehaene's account of cognitive (neuro)-psychology, one of the specialized mental organs of the brain is a primitive number processor which prefigures, without corresponding exactly to it, the arithmetic taught in school. This "organ" is present in species other than humans and would provide animals and humans with direct intuition of what a number means (Dehaene [37]).

Although this organ, this "sense of numbers", has limited scope and functions and handles only approximately integers beyond the very first ones, its very existence shows that numbers have a physical and pre-linguistic dimension: they are apprehended, in animals, outside of discursive patterns, and it is probable that, even in adult humans, this sense continues to play a role in the apprehension of quantities.

13.5   Didactics

One of the main interests of these remarks lies in their implications for the teaching of arithmetic and, more generally, science. For young children, this idea is well highlighted and illustrated by numerous examples in *The Number Sense* (Dehaene [37]). Given that the actual learning of numbering is a manifold process in which the "number sense" plays an important role, it is illusory to look for a normative view of number learning based on the post-Fregean conception of number. Yet this is indeed what the didactician Piaget (1896–1980) wanted to do, leading to rather surrealistic debates on the role of set theory in the acquisition of fundamental notions of arithmetic.

Of course, very young children do not and cannot acquire the concept of number: there is a long way to go from individual concepts (mum, dad, teddy...) to material concepts (toy, horse...), then theoretical concepts (numbers...). The mistake would be to believe that the teacher can bypass these phases of constitution. This remark is still valid at advanced stages of education: in the 1950s and 1960s, geometry was taught without figures to students from the École normale supérieure by the Bourbaki school (Cartier [27])! This led to an impoverishment of intuition among students confronted with such methods. Personally, having been educated in mathematics at a time when structuralism was still influential and continued to favour certain educational standards, I often came to understand, at the cost of detours, chance or complex thought processes, that the theoretical notions I had been taught had a very simple, very intuitive concrete meaning. Paradoxically, this effort to 'deconstruct' a theory, perhaps because it was more uncertain and delicate, less conventional than learning theoretical content, gave me real intellectual pleasure.

In addition to all this, it should be added that the mask of abstraction is sometimes used by teachers but also in the world of research and advanced mathematics to give more weight, credit and respectability to the results presented. Another personal testimony (it is understandable that I have to avoid going into detail, especially since the example is worth more by its generality than its specific content): some time ago I attended a conference where several presentations were given on a subject that I knew only marginally. Out of curiosity rather than to try to dominate it, I decided to browse through the existing literature. I read, among other things, an article published in one of the very best mathematical journals—a difficult article, not very explicit on many points and referring to very abstract notions; in short, a rather inaccessible article for the neophyte. Later during the conference, I had a discussion with one of the best specialists on the subject, who advised me very strongly to read the PhD thesis from which this article was taken. The thesis was easy to read, pleasant and deep, with motivating and illuminating examples and calculations—a beautiful mathematical work, very stimulating, which left many questions open, where the article that took up the content left little room for the work of the imagination. The explanation of the rather absurd and collectively counterproductive path that had led from the clarity of the thesis presentation to rather jargonish choices of exposition seems to be due to the caudine forks of academic publishing, which would not have been satisfied with a too direct, too elementary presentation of the results. Every mathematician will have had similar experiences. Now, abstraction is beautiful only when it is useful, producing meaning and generality, and, even when it is useful, it is clumsy and goes against the interest of science to conceal its moments of constitution and the simple examples that give it flesh and body.

13.6   Complexity of Numbers

Another occurrence of psychological phenomena in the elementary theory of numbers, more subtle or at least less classical than those we have reported so far, is the idea of complexity. When we talk about whole numbers, we almost always have "small numbers" in mind, and if we ask anyone to give us a whole number at random, that number will almost always be "small". This is of course because the sequence of integers is infinite, but also, more profoundly, because we look at this sequence with a psychological bias: we are most often dealing with integers with one, two or three digits, and will therefore put more "weight" on them when it comes to saying a number at random.

There is therefore a gap between mathematical theory and the human and computer use of numbers where, in practice, all calculations are made on an initial segment of the set of positive integers. Similarly, suppose we flip a coin, with H being "heads" and T being "tails". We will find the sequence of results HTTHTHHHHHHHHHH normal, but we will be immediately convinced that the game is biased if the sequence is HHHHHHHHHHHHHHH even though, statistically, the two draws are equiprobable (they have the same probability of occurrence, if the game is not biased).

We are thus able to analyze in a spontaneous and non-thematic way statements of an arithmetic nature: it is not necessary to know the probabilities to suspect that the second case is an indication of cheating. These ideas have been the subject of a mathematical theory, that of complexity;[1] an interesting example of a situation where the thematization of fairly elementary intuitions linked to our most spontaneous and ordinary cognitive practices has given rise to a deep and active field of research.

13.7   Malleability of the Brain

While there is ultimately no doubt that our thinking is limited by our cerebral capacities, determining the extent of its possibilities and capacities remains difficult and conditioned by the field of our experience, both individual and collective. The twentieth mathematical century has thus developed a whole set of techniques such as structural algebra, algebraic topology, algebraic combinatorics and category theory that profoundly modify the very idea that mathematicians may have of their discipline, its springs and its scope. The mathematician's brain and thinking, after a phase of learning (individual and collective), adapts to these new tools which become as natural, obvious and intuitive to him as the use of numbers or geometric figures can be to children.

---

[1] See Bienvenu and Hayrup [11].

This extreme malleability of the brain, its ability to integrate and domesticate very complex thought patterns, is always surprising, including in its historical dimension: calculations on fractions or decimals performed today by 10-year-olds were reserved for an elite in the Middle Ages, a phenomenon that the evolution of notation and calculation techniques does not fully explain.

Let us depart for a moment from arithmetic: a concept as primitive as that of point, whose obviousness rivals that of number, has thus seen over the last century the perception and intellection that mathematicians had of it radically changed. Under the influence of algebraic geometry in particular, geometers began to think of a point as characterized by the set of functions that vanish at it. As this set has the structure of a maximal ideal (a concept born in algebra, with the works of Kummer, Dedekind, then those of the German algebraic school, with in particular E. Noether), the concept of maximal ideal has been identified little by little, in modern geometry, with that of point, with all the consequences that can be drawn from this evolution. They involve the widening of the very notion of point and, concomitantly, of the geometrical notions associated with it: infinitesimal neighbourhood, existence of new topologies adapted to the new algebraic view of geometry—Zariski topology, and other exotic topologies at the root of the most recent and most spectacular results within the confines of algebraic topology and arithmetic geometry.[2] These new ideas have a feedback on the intuitive perception that mathematicians have of a point; the view that such a concept would have been frozen once and for all with the choice of a privileged axiomatic appears in passing devoid of any historical legitimacy.

These observations are not without incidence on the way in which mathematical intuition is to be understood—for example, the faculty of forming representations of more or less complex ideal objects (hexagon, group, algebra...). Mathematical objects are often intuitive to us when they can be associated with familiar representations or ideas: for example, geometric figures and even abstract concepts such as the one of group, to which it is easy to associate a certain number of intuitive representations (rotations, translations, permutations...). The faculty of intuition would thus result first of all from the empathy of certain idealities to our cerebral structure. However, mathematical experience encourages us to relativize this naive conception. The malleability of the brain and its adaptability lead us to believe that complex and new neuronal connections are capable of forming each time a new field of mathematical research opens up. Whatever the complexity, however off-putting the definitions, however unintuitive the theory may seem at first glance, a few mathematicians will master it, followed soon, if the subject is worthwhile, by many students for whom these notions, which they discover already formed and sedimented, will more or less be taken for granted. It seems very difficult today to predict how far the cognitive sciences, whose field of action remains for the moment limited to fairly simple processes, will be able to go in the empirical description and physiological analysis of this type of phenomena, typical of modern mathematics.

---

[2]Dieudonné [42, 43].

Chapter 14
Phenomenology of Numbers

> *The time in which we live is so full of its apparent clarity that it has ceased to feel the difficulties, let alone try to solve them. It has only one title of glory which must be fully preserved: it is to have succeeded in technically developing calculation with a magnitude and a state of completion that previous eras had never known. It is true, it will be realized, that there is also precisely in this perfection of modern arithmetic one of the main sources of error. Today's arithmetic is an unsurpassable art for the breadth of its rules and the possibility of trusting them in practice; but it is not a science, if we understand by science a system of knowledge.*
> *Husserl [64]*

The end of the nineteenth century and the beginning of the twentieth century marked a turning point in the way mathematicians approach their discipline. Even more than the attempts to refound the corpus on the basis of set theory, the successes of the axiomatic method challenged the traditional conception of the meaning of mathematical statements, which had hitherto been largely based on a spontaneous belief in an adequacy between these statements and their empirical and physical correlates.

These questions were, admittedly, not entirely new: the emergence of non-Euclidean geometries had thus begun to shake up a certain ontological naivety,[1] but these geometries remained to a large extent compatible with our fundamental intuition of space (which is not necessarily Euclidean, as the debates of the time showed). The Hilbertian axiomatization of geometry, on which most of the debates crystallized, went further by cutting the privileged link, which non-Euclidean geometries had not broken, between geometry and the intuition of space.

---

[1] See Boi [13].

F. Patras, *The Essence of Numbers*, Lecture Notes in Mathematics 2278, https://doi.org/10.1007/978-3-030-56700-2_14

A discourse[2] by the great geometer Felix Klein, delivered in Vienna in 1894 in honour of Riemann, illustrates well the concerns and state of mind of the time. For Klein, Riemannian mathematics, one of the greatest achievements of nineteenth-century mathematics, rested on two foundations: attention to physical problems, and a guiding idea, "to grasp the properties of things according to their existence in the infinitely small". However, Klein is careful to distinguish the question of the origin of theories of the logical organization by axioms, since the genetic and formal moments of a mathematical theory should not be opposed but considered complementary. The discourse clarifies this articulation:

> It is not indifferent, in the search for and discovery of mathematical laws, to attribute or not a determined meaning to the symbols with which one operates. Indeed, the concrete presentation provides us with the link between ideas that leads us forward [...]. The results that derive from research in pure mathematics are, however, above any kind of particularization. It is a logical general scheme, a system whose particular content is not indifferent, as this content can be chosen in various ways.

Generality and validity of a logical scheme on the one hand, role and non-indifference of meaning on the other. Klein rejects a dichotomy of the symbolic and intuitive thinking. Hilbert will not express himself differently at the beginning of his *Foundations of Geometry* [60] or in his later work.

14.1 The Problem of Origins

It is in this particular context of profound questioning of the very content of mathematics that the work of Husserl, creator of Phenomenology, takes shape, a philosophical and epistemological technique to which this chapter and the next will be largely devoted.

Husserl, when he published his first major work, the *Philosophy of Arithmetic* [71], in 1891, read Frege's *Foundations of Arithmetic*,[3] and opposed the idea that a formal and logical definition of numbers could be sufficient to found arithmetic. Beyond the problem of numbers, the quest of philosophy of mathematics cannot stop at the determination of systems of axioms, at the logical-formal dimension of the discipline. It must also deal with everything that is below them and guarantees its very possibility, both from the point of view of the historical processes of constitution and of the logic internal to the development of science:

> You can see where Frege is going with this... A foundation of arithmetic on a series of formal definitions from which all the theorems of this science can be derived in a purely syllogistic manner, this is Frege's ideal [...]. One can only define what is composed in a logical way. As soon as we encounter the ultimate, elementary concepts, all defining activity

---

[2] Available for example in the French translation of Riemann's Complete Works, Paris, Gauthier-Villars, 1898.

[3] On Husserl and Frege, we refer to Brisart [19].

comes to an end. No one can define concepts like quality, intensity, place, time, etc. And the same is true for elementary relationships and the concepts that are formed on them. Equality, analogy, gradation, the whole and the part, quantity and unity, etc., are concepts that are not at all susceptible of a logical-formal definition. What can be done in such cases is only this: to show the concrete phenomena from or in the midst of which they are abstracted, and to clarify the nature of this abstraction process.

Husserl [71, pp. 145–146]

The problem that Husserl thus poses, in a way that will become more refined after the *Philosophy of Arithmetic*, in the mature writings such as *The Origin of Geometry* [69], is the one of origins. Understanding mathematics cannot be limited to drawing up the state of constituted mathematics (which is, moreover, the task of mathematicians themselves). It is necessary to go beyond this and look for the very reasons for the emergence of concepts such as that of number. It will appear later, in texts such as *Formale und Transzendentale Logik* [72], that this task is not without incidence on parts of mathematical philosophy that go beyond the framework of foundation problems, since it is the whole of mathematical activity that is finally likely to be questioned through the prism of the modalities of emergence and constitution of idealities.

14.2 Three Points of View

Husserl developed, over the years, three points of view on arithmetic, which can provisionally be described as psychological, symbolic and algebraic (this summary classification will be further clarified and explained in the following paragraphs). These points of view are all in germ in the *Philosophy of Arithmetic* and correspond to three different and complementary moments in the phenomenological analysis of number.

Husserl's analysis has evolved during his long career, and he was for example quite quick to reject the idea that psychology could legitimately support a foundation of the idea of number (a thesis he defended in the *Philosophy of Arithmetic* but abandoned as soon as in the *Logical Investigations* [73]). Since the subject of this chapter is not historical or exegetical, we will pass in silence over the detailed presentation of the different moments of the Husserlian work and the variations they induce on his analysis of numbers, to insist rather on the underlying guidelines which, in the end, have not moved much from the beginnings to maturity. In particular, we will use the phenomenological vocabulary quite freely, including when analyzing works that predate the development of phenomenology *stricto sensu* and its techniques.

14.3   The *Philosophy of Arithmetic*

The *Philosophy of Arithmetic* is from 1891: a text from his youth, in which Husserl did not yet have the tools of phenomenology, and neither familiarity with Hilbertian ideas and the problems linked to the emergence of the axiomatic method, a familiarity that he would conquer only in the last years of the nineteenth century. However, most of the major issues of the Husserlian theory of knowledge are already there: the relationship between intuition and symbolic knowledge; the autonomy of the formal (of the domain of signs, more precisely) with regard to the activities of consciousness; the passage from representations and their internal structure to associated concepts. All these elements, immediately identifiable in the later work, are, however, only present there in watermark. The discovery of Hilbert was then to lead Husserl towards a consideration of the possibilities offered by the axiomatic method.

Prior to this shift of Husserl's thinking and interests towards the logical-formal domain, the text's main interest for our purposes is obviously that, unlike the later major texts, the focus is on arithmetic. According to the still rather naive approach of this early work: "We can consider that it is not in itself a blameworthy thing that mathematicians, instead of giving a logical definition of the number concepts 'describe the way in which one arrives at these concepts'; however these descriptions would have to be correct, and also fulfilling their purpose."

The method chosen by Husserl refers first of all to psychology: the concepts of numbers are formed from contents of representations that undergo two stages of abstraction. The first is that of collectivization: in order for me to affirm that there are five apples on this table, I must first organize these apples into a collectivity, a multiplicity. This is accomplished spontaneously by our consciousness, but it is not something obvious or spontaneous (apples do not spontaneously constitute themselves in a collection, there must be a psychic act). The second step, which allows us to pass from multiplicity (the apples constituted in a distinct totality) to number (the number 5) is more complex.[4] There must be a shift in psychological interest, with attention having to shift from the multiplicity of individualized objects to a multiplicity where the contents are understood in the mode of the simple "something". The thesis corresponds to the classical conception of a number as a "collection of units", where the units are the individual objects but where individuality is put in brackets. This Husserlian conception of number would thus be nothing but a variant of classical theses, but for the way in which these successive displacements (multiplicity, number) of psychological interest are analyzed. By seeking to think about the cognitive legalities proper to this displacement, Husserl will understand that the creation of idealities has its own logic; we will come back to this.

---

[4]It is well analyzed by R. Brisard in *"Le problème de l'abstraction en mathématiques"* (Brisart [19]), which we follow here.

14.4   Proper Representations and Intuitions

The first part of the *Philosophy of Arithmetic* thus deals from a predominantly psychological point of view with questions relating to the analysis of the concepts of quantity, unit and number, as long as they are given in a proper way and not in an indirect symbolization, the second part dealing with this second, symbolic mode of accessing numbers.

The Husserlian project is best understood in the light of slightly posterior texts: Husserl explains the notion of "proper" and the idea of intuition in a remarkable text from 1894, the *Psychological Studies for Elementary Logic* [66]: "Intuition is not a 'representation' in the improper sense of a simple replacement by pieces, images, signs, etc., nor a simple determination by distinctive marks—means by which the represented is absolutely not, in reality, presented before us. It is a representation in a more proper sense, which effectively presents before us its object, in such a way that it is itself the substratum of psychic activity." (Husserl [66, p. 170])

The word representation, widely used in phenomenology, is not without ambiguities, often commented on in the philosophical tradition;[5] "representation" will designate in what follows an experience of consciousness directed towards an object, whether this object is actually present and perceived, ideal (a number, a figure, an idea) or associated with a memory or an imaginary object (the centaur, a "square circle").

The adequacy of the representation to its object is not self-evident: our perception can deceive us, but what is true of sensations is still true of ideal "perceptions". This was the purpose of the first Platonic dialogues: we rarely know what exactly the concepts we use (good, justice...) mean. These remarks continue to be valid in part in the scientific field. We can be mistaken in our apprehension of mathematical objects. In elementary geometry, a badly chosen figure will wrongly convince us that a certain angle is a right angle or that a certain ellipse is a circle. In algebra, we will think we are dealing with a group without having verified that a certain relation (associativity, for example) is satisfied. These errors are frequent and are part of the mathematician's daily life. They are moreover often profitable as soon as they allow us to put our finger on the key properties of the problems studied: it is often the gap between our naive perception of a problem or a mathematical object and its actual properties, more complex, that gives access to their authentic understanding.

A representation in the truest sense of the word, that is, where there is an adequacy of the representation to its object, is an *intuition*. Intuition intended in this sense is therefore constitutive of knowledge since it is it that makes it possible to speak of understanding. However, not everything is intuitionable, and this structural limit of our cognitive faculties has important consequences for the philosophy of space and number:

---

[5]Notice that we follow here the French phenomenological tradition: the meaning and use of terms such as "representation" may vary from one language and hermeneutical tradition to another.

> Figures and geometrical relationships are absolutely not intuitionable if we have to prove
> right those who are afraid of wrongly attributing ideal properties, which perceptions
> cannot show in space, to the corresponding products of the imagination. The objectives
> of the idealizing processes, therefore conceptual, are *eo ipso* non-intuitive. The figures
> and relations actually intuitioned "represent us" by virtue of certain analogies (infinitely
> coarse, measured to the ideal of definition), the figures and geometrical relations properly
> intentioned, often even substitute themselves for them in the living geometrical thinking.
> Anyone who has become aware of this will immediately refuse to speak of an intuition of
> geometric abstracta, as this is an improper way of speaking.[6]
>
> Husserl [66, p. 173]

If the analysis of intuition is more or less self-evident for geometry, as its
anchoring in the perception of space is so strong and necessary, the theory of number
also lends itself to it:

> The mathematician sometimes says: "The sign a will represent an arbitrary number, it will
> represent the root of this equation", instead of: it will designate it. One will hardly be
> inclined to call this representation an intuition, obviously for reasons similar to the cases
> discussed above. According to Kant, no doubt, the sign drawn by the arithmetician would be
> a construction in intuition [...]. But it is absolutely inadmissible to consider the writing of
> the arithmetic sign as a construction. The sign and that which is designated are here totally
> foreign in content and are linked only by association. The sign therefore does not intuitively
> express what is thought, it merely refers to it. Moreover, in the present case of arithmetic,
> what is designated is almost always something that cannot be subject of an intuition at all.
>
> Husserl [66, p. 174]

It should be noted in passing that these theses are in line with the findings of
cognitive neuropsychology: numbers are conceived in the brain in very different
modes, ranging from an immediate and almost perceptive mode when it comes to
the very first numbers to complex modes linked to language and symbolism.

This remark being made, it is necessary, on the basis of these strong theses on
mathematical intuition and its limits, to understand Husserl's approach, which is not,
as is too often believed, to aim at a pure and simple reduction of all knowledge to
a rather naive form of originarity of knowledge linked to perception and the world
of life. Very early in his work, a dichotomy was established between theoretical
knowledge, whose relationship to intuition is complex and structurally limited, and
the problem of genesis, of the origin of knowledge, where intuition finds its full
place. When intuition does not have access to the "things themselves", other modalities of access replace it: "[There is] a separation between the 'representations' that
are intuitions and those that are not. Some psychic experiences, generally called
'representations', have the particularity of not containing within themselves their
'objects' as immanent contents (thus present to the consciousness), but simply to
give access to them by intentional consciousness in a certain way..."

---

[6]Husserl questions here the very idea of intuition in a geometrical context, emphasizing that the
relationship between abstract entities (even simple ones such as, say, the notion of triangle) and the
representations that ground our grasping of them relies on highly complex processes. These ideas
echo the treatment of imagination in Kant.

14.5 Back to One

The detail of the phenomenological analysis can only be understood through examples. Husserl returns for example to the problem of the unit, departing from the classical approach to the definition of a number as a quantity of units in the manner of Locke (which was already based on a psychological analysis):

> Amongst all the ideas we have, as there is none suggested to the mind by more ways, so there is none more simple, than that of unity, or one: it has no shadow of variety or composition in it: every object our senses are employed about; every idea in our understandings; every thought of our minds, brings this idea along with it [...]. By repeating this idea in our minds, and adding the repetitions together, we come by the complex ideas of the modes of it. Thus, by adding one to one, we have the complex idea of a couple; by putting twelve units together, we have the complex idea of a dozen; and so of a score, or a million, or any other number.
>
> Locke [82, book II chap. 16]
> Quoted by Husserl [71, p. 139]

The problem with this kind of definition of number is that it treats "units either as concrete contents, simply by sticking to names or signs of writing, or (which is the general rule) as abstract, positive partial contents that could be detached in isolation and joined together in collections to form multiplicities". But: "Any enumeration would be totally meaningless if the sign 1 or the word one did not have the meaning corresponding to the concept of one, that is, if it did not designate the abstract process that removes the limitation of the singular object by transforming it into a simple something or a one [...]. By simply sticking to enumeration as an external mechanical process, we have completely forgotten to see the logical thought content that gives it justification and value in our entire mental life." (Husserl [71, p. 141]) In the end, "the difficulty lies in the phenomena, in the correctness of their description, analysis and interpretation; only by referring to them can one come to see what is the essence of the concepts of number".

14.6 The Transcendental Point of View

Husserl's distantiation from Locke and the other proponents of a conception of number as a collection of units is thus first of all a question of method: the epistemology of number cannot simply describe the conceptual contents, it must look at the structure of intuition and intentionality, because it is in this structure that the constitution of numbers is at stake.

That such a structure exists and is identifiable, that it emerges from the way phenomenological analysis looks at the relationship between intentional consciousness and its objects (the so-called duality between noesis and noema or between noetic contents of intentional acts and their noematic correlates), that it is finally capable of justifying the constitution of scientific idealities, are all powerful theses that will emerge from later Husserlian works. Nevertheless, these ideas are in germ in the *Philosophy of Arithmetic*, whose commentators have underlined the key and prefiguring role for all Husserlian thought.[7]

The Husserlian understanding of number, as has been said, is played out in two moments. The first, the formation of a collection, makes multiplicities emerge in a mode quite close to the Cantorian conception. The second, from which number is born, is linked to the general concept of "something" which translates the fact that we are capable of forgetting what is individual in the members of a multiplicity in order to bring them under a general and undifferentiated concept that will allow us to consider them as so many ones, as many units.

> We have to characterize this concept more precisely in terms of its content and formation [...]. The concept of something cannot, of course, be acquired by any comparison whatsoever between the content of all physical or psychic objects. Such a comparison would simply be fruitless. The "something" is precisely not an abstract partial content [...]. It obviously owes its formation to the reflection on the psychic act of representation to which is given precisely as content any given object.
>
> Husserl [71, p. 86]

Thus, "this concept [of the something] is not obtained by simple abstraction from the contents, it rather owes its formation to the reflection on the act of representation of the contents [...]. This suggests that [from the *Philosophy of Arithmetic* onwards] abstract processes are susceptible to a double approach, since, unlike material abstraction in the sense of the simple operation of reducing the specific properties of the contents, allowing them to be understood together, Husserl already envisages formal abstraction in the sense of a reflection on certain acts or certain intentional aims by means of which the something in general, multiplicity as pure form, etc., become a modality of donation of the contents of representation". [8]

"This abstraction of a new type is only possible through an operation of a radically different type: a reflection, which here is what we could call intentional reflection, a return to the aims of consciousness as such, retained in their pure formality, their general character of aiming at the object. Such is the operator of passage to the *something*, which no natural separation of content could have isolated in the consciousness. At some point, the consciousness must turn to itself."[9]

The reflexive return to intentionality thus makes possible the eidetic constitution of universals, according to a transcendental mode close in many aspects to Kantian philosophy and which will later become, in the Husserlian work, typical of mathematical creation.

---

[7]We refer to Brisart [19] and in particular to the contributions of J. Benoist, R. Brisart, and the author.

[8]R. Brisart in [19].

[9]J. Benoist in [19].

14.7   The Problem of Symbolism

The second part of the *Philosophy of Arithmetic* raises different but equally delicate questions, since Husserl seeks to account for the validity of symbolic writing and calculation, its effective scope and even its primacy in the daily use of numbers.[10] It does not solve the problem of the passage from the foundation of concepts on acts of thought to the legitimacy of a systematic use of modes of calculation based on signs, and its study must be, from this point of view, completed by that of a Husserlian text of 1890, "On the Logic of Signs" [65].

After stating the relativity of the concepts of relation and object, the text continues with the remark, which is decisive for the economy of Husserlian thought, that improper representations[11] are "the foundation of our practical activity of ordinary judgment".

> In the unfolding of a rapid flow of thoughts, signs exercise substitution without us being aware. We believe we are operating with the actual concepts. But even if we are forced to reflect and notice the true state of affairs, as when we suddenly become uncertain and start to meditate on the meaning of a word, we are usually satisfied with mere substitutes.
>
> We do not notice that we are operating with substitutes instead of the full concept. So it is the same for our judgments as for our representations; instead of our own judgments, we have symbolic judgments, but we do not notice that they are.
>
> Husserl [65]

Thus, "it is therefore certain that it is not logical motives, that is, motives of knowledge, that guide us in the practical activity of judgment, but blind psychological laws". Everything happens as if the ordinary flow of our reflections was carried by purely signitive structures, and this with sufficient validity for us to be satisfied with this substitute for authentic thought, that is to say, for knowledge through the adequacy of a conceptual aim and an intuition. Husserl thus comes to envisage a specific form of legality for symbolic reasoning processes.

These remarks are obviously surprising, especially if one compares them with the later *Psychological Studies for Elementary Logic* [66], where he seems perplexed as to the significance for the theory of knowledge of the normative power of the sign:

> I in no way wish to claim that one cannot make any significant progress in the logical understanding of the correctness of symbolic thought (at first naturally of mathematical thought) without a thorough understanding of the essence of the elementary processes of intuition and re-presentation that mediate it everywhere; but one does not, however, acquire a complete and effectively satisfactory understanding of these logical processes (as of all others) without such an understanding.

The ideas developed in the "Logic of Signs" are, for the most part, consistent with the developments of the second part of the *Philosophy of Arithmetic*. They apply to

---

[10]In this paragraph and the preceding ones, we repeat some of the analyses of F. Patras, "*Le fondement de l'arithmétique*", in [19]. We refer to this article for more details.

[11]In this case those that take place through the mediation of signs, to which we will return in the following section.

lived arithmetic, where we most often operate with numbers without regard to the corresponding arithmetic concepts. This theory of signs leads to two fundamental questions.

The first relates to the reasons for autonomy, and is formulated by Husserl: "If a typical process of judgment, although not guided by reasons of knowledge, nevertheless leads to just results, one must seek and find in its internal constitution the reason why it is capable of producing truth." (Husserl [65]) Husserl's answer to this question is very close, in its argumentation and springs, to what mathematicians at the beginning of the century would have called the axiomatic method.[12] When a reasoning or a theory presents forms of invariance, its structure is valid independently of variable terms. The form of the reasoning is sufficient to guarantee its validity, and "knowledge of such a situation makes us capable of putting in place of effective reasoning a symbolic reasoning, in a conscious manner and for logical reasons" (Husserl [65]).

The second is the possibility of reactivating meanings in symbolic reasoning and statements. This is the question that will still preoccupy Husserl in the *Krisis* [70] or the *Origin of Geometry* [69], this reactivation taking the form of recourse to a system of originary intuitions.

14.8 Improper Representations

Understanding how such ideas apply to lived arithmetic is instructive and provides a radically new perspective on the nature of numbers.[13] Let us recall the observation made at the end of the psychological and intentional analysis of multiplicities and the formation of the concept of number from the "something": all these moments of eidetic constitution do nothing to justify the power of arithmetic, its autonomy with regard to our representations of collectivities in the daily practice of arithmetic, or even the way we operate with large numbers.

In fact, it seems that improper representations of numbers, that is, those which do not correspond to intuitive moments, are the rule. Husserl therefore embarks on a systematic study of symbolic representations:

> We will first explain in a few words the difference between symbolic representations and proper representations, which is fundamental for the developments that will follow. A symbolic or improper representation is, as the word already says, a representation by signs. If a content is not given to us directly as what it is, but only indirectly through signs that univocally characterize it, then, instead of a proper representation, we have a symbolic

---

[12]The term "axiomatic" was then to take on a narrower and more specifically logical meaning.

[13]Here we follow partly Patras [91], to which we refer for more details.

representation of it [...]. Any description of an intuitive object tends to replace the actual representation of this object by a representation of signs that stands for it.[14]

Husserl [71]

To understand the exact nature of the problem, it should be noted that the field of symbolic activity in mathematics is largely independent of formalization or axiomatics: children who operate on number signs by doing elementary operations in a mechanical mode have no idea of formalized mathematics! The symbolic representations, which Husserl mentions, and on which he ultimately bases his theory of numeration, therefore have an original content. We will see, in the next chapter, how to reconcile them with the phenomenological point of view, an intentional approach and fairly recent mathematical ideas.

14.9 Horizon Structures

Let us return for a moment to the second question raised by symbolism—that of a reactivation of meaning. In everyday mathematical activity, thought moves in the game of formulas and the meaning of objects are rarely involved. Or they are involved in an improper way, for example through the typical properties of this or that object, without it being necessary to take the thought back to the object itself when stating these properties.

Thus, in a calculation of integrals, it is generally completely useless to have in mind the meaning of the integral operators: it is enough to know the rules they obey (integration by parts or Fubini's formulas...). In opposition to this indifference to the meanings of mathematical objects in their ordinary frequentation, creative reasoning and discovery are based on a principle of horizon. Although we relate to the objects in an improper mode, it is possible to change this mode each time. Each object thus carries with it a horizon of structures and intuitions, of theoretical possibilities.[15] In other words, the mathematician is able to orient his thinking to cope with the problems he or she faces, which implies that concepts are not fixed in an exclusively symbolic system. The possibility of reactivating original meanings is part of this horizon structure: even when we operate in a purely signitive way on numbers, it is essential that we can, when we wish or when the argumentation requires it, interpret our calculations in intuitive terms. But it is also essential that we can abstract ourselves from this interpretation in order to freely give other meanings to the numbers.

---

[14]Symbolic algebra, where we operate blindly on signs using mechanical rules whose meaning need not be reactivated, or the use of computers to handle symbolic computations are typical examples of such phenomena that pervade mathematical practice.

[15]The notion of horizon plays a central role in Husserl's phenomenology, see [67, Sect. 19], Actuality and potentiality of intentional life and ff.

A great discovery, essentially empirical and not theorized, of twentieth-century mathematics is that this horizon structure of mathematical objects can be organized.[16] To a certain extent, it is possible to describe the way intentionality varies around an object, to consider it according to various modes of existence or meanings (the method of eidetic variation, in the language of the *Ideen* (Husserl [68])). In the case of numbers, this will mean that they can be considered as properties of sets, as associated with ordered structures, as representatives of classes of integers modulo a prime number p, or even as systems of symbols. All these processes are legitimate, as long as none of them claims an exclusivity in the order of the foundations, and each one participates, in fact, in the horizon structure of numbers as we know and use them.

---

[16]In his own way, and with his own epistemological prejudices, Bourbaki says much the same thing in "*L'architecture des Mathématiques*" [18].

Chapter 15
Universal Phenomena, Algebra, Categories

Contemporary mathematics is the result of a long process and, as such, has many hermetic or esoteric features that make it extremely difficult to access. This ever-increasing complexity is, however, accompanied by a deepened and renewed understanding of the mechanisms of scientific knowledge and a more refined perception of their fundamental concepts. This qualitative progress, this deeper understanding of the springs of mathematics is not always obvious to detect in everyday life. In fact, at first sight, the twentieth century seems to have accomplished the programme that was at one time that of logical positivism: a systematic abandonment of so-called "metaphysical" questions in favour of experimental and hypothetico-deductive methods, leaving little room for reflection on science that would not be a mere commentary on scientific progress.

Some of the questions perceived as intrinsically "metaphysical" by the philosophical tradition, such as that of the nature of mathematical objects, first and foremost numbers, are nevertheless likely today to receive a new meaning as a result of scientific progress. One of the reasons for this is that mathematicians today are accustomed to working with many more ideal objects than a century ago: to classical numbers and geometric figures have gradually been added algebraic entities (groups, algebras...), esoteric geometric figures and spaces (non-Euclidean geometry,[1] the spaces of general relativity, some more abstract spaces where

---

[1]The spaces of non-Euclidean or Riemannian geometries (such as those of general relativity) are "curved" spaces: to define measurements (distances, volumes, etc.), it is necessary to take into account, from a physical point of view, force fields (gravitational, electromagnetic, etc.). The geometrical intuition associated with these spaces remains fairly close to intuition in Euclidean geometry, at least at a local level. The abstract spaces of contemporary geometry or topology are much more difficult to conceive. For most of them, it is impossible to have a direct intuition as they are defined by their formal and algebraic properties rather than through explicit constructions. The traditional dividing line between the intuitive character of geometry and the formal character of algebra no longer makes much sense. The usual epistemological reductions, which dissociate

© The Editor(s) (if applicable) and The Author(s), under exclusive license to Springer Nature Switzerland AG 2020
F. Patras, *The Essence of Numbers*, Lecture Notes in Mathematics 2278,
https://doi.org/10.1007/978-3-030-56700-2_15

any form of spatial intuition proves inoperative, if not metaphorical), complex combinatorial structures[2]. . . .

This enrichment of the field of objects and theories has profound consequences. In algebra, for example, the whole of the nineteenth century and the beginning of the twentieth century had gradually put forward the idea of the pre-eminence of formal structures over the anchoring of algebraic theories in particular object domains. Thus, an algebraic entity such as a group or an algebra, which historically appeared rather through operations (so that the first groups were naturally defined as groups of operations on spaces or sets), has been defined, since the emergence of "modern algebra", by giving a set and laws of composition on this set such as addition, multiplication or more general laws. However, in contemporary algebra, these laws are sometimes so complex that they do not lend themselves well to symbolic manipulations: one then chooses, in order to understand and study them, to associate them with geometric representations[3] and to work with these geometric representations (diagrams, graphs and mathematical trees) rather than with the symbolic representations (in terms of operators) familiar in classical algebra treatises. This development of a combinatorial geometry of algebraic structures raises a number of questions about the nature and functioning of algebraic intuition, partially challenging the naive imagery of an unconditional primacy of symbolic and formal operations that was in vogue with "modern mathematics" from the Second World War to the 1980s.

Thus, in the movement of science, major categories of thought are undergoing structural transformations. The distinction between objects, forms, transformations and concepts is fading; the distinction between theorems and proofs lost its validity through research in theoretical computer science, where proofs became objects of study in their own right.

In this last chapter, we will examine the consequences of some of these transformations, mostly those related to the progress of algebra since the end of the nineteenth century, progress that gives a renewed tone and content to the symbolic approach to numbers that had first emerged with Leibniz, Cauchy, Grassmann and the English school.

---

algebraic intuition from combinatorial or geometric intuition, thus lose their legitimacy. See for example Dieudonné [44].

[2]Partially ordered sets that generalize the total orders considered by Cantor, graphs, trees, combinatorics of words. . .

[3]The links between algebraic theories and combinatorial and geometric objects were developed by the Montréal school of algebraic combinatorics, with the theory of structure species; see Bergeron et al. [10]. These links have been taken up and developed in various forms very recently and play an important role in contemporary research, from combinatorial to algebraic topology, from particle physics to the theory of differential equations. The work of Maxim Kontsevich, who was awarded the Fields Medal in 1998, is a good illustration of this trend.

15.1 Numbers as Invariants

Let us start with a methodological remark. Numbers are, without doubt, among the very first mathematical objects encountered, and numbering implicitly underlies many elementary cognitive processes. It is therefore legitimate to approach the problem of the constitution of numbers in a genetic (how their concept emerges) and logical (what relationship do they have with the theory of pure forms of knowledge) mode, in the manner of Frege and a whole epistemological tradition. However, we can also legitimately approach numbers as a first example of more complex mathematical notions. They receive an insight in return that is not without interest, including even from the classical point of view of the elementary constitution processes of the idea of number.

The theory of invariants and Kummer's theory of ideal numbers,[4] very much in vogue at the end of the nineteenth century, are part of this programme of re-examination of the idea of number. Let us first come back, with Kronecker, to the question of the primacy between the concepts of ordinal and cardinal numbers, a question which was generally decided in favour of one or the other depending on whether one chooses a recursive, dynamic approach to numbering (like Dedekind) or a set-theoretic approach (like Frege):

> The natural starting point for the development of the concept of number, I find it in ordinal numbers. With them we have a reserve of certain designations, ordered in a fixed successive sequence, which we can assign to a group of objects that are different and at the same time distinguishable. We place the totality of the designations thus used in the concept of the "numbering of the objects" that make up the group, and we unequivocally link the expression of this concept to the last of the designations used, since their succession is determined in a fixed manner. Thus, for example, in the group of letters *(a, b, c, d, e)*, to the letter *a* can be attributed the designation of "first", to the letter *b* of "second", etc., and finally to the letter *e* of "fifth". The totality of the ordinal numbers thus used or the "numbering" of the letters *a, b, c, d, e* can therefore, by being attached to the last of the ordinal numbers used, be designated by the number "five".[5]

Kronecker thus favours ordinality, but for reasons that appear to be related to an original understanding of the concept of number, as explained in an 1891 lecture, *On the Concept of Number in Mathematics* [80]. As he notes in the introduction, number can be approached not only from a genetic point of view (that of the mathematical knowledge of children or of individuals lacking mathematical knowledge), but also from the point of view of those "who already possess sufficient knowledge of mathematics" because "only in this way is it possible to bring to the perception of the concept in all its clarity and acuteness, something which is never possible through a simple philosophical definition".

---

[4]On Kummer's ideal numbers and the role they played, especially in Dedekind, we refer to the analysis of Boniface [15, chap. III].

[5]Quoted by E. Husserl [71, p. 197].

Kronecker's idea[6] is to associate the construction of numbers to the theory of invariants, whose origins he credits to Gauss and Lagrange. At its beginnings, the theory dealt with quadratic forms $Q(x, y) = ax^2 + bxy + cy^2$. Two quadratic forms Q and Q' are said to be equivalent if one can be transformed to the other by a change of variables.[7] The study of one is thus equivalent to that of the other. The problem of the theory of invariants is to characterize the functions f of Q (i.e. of a, b and c) such that $f(Q) = f(Q')$ when Q is equivalent to Q'; $f(Q)$ is then called an invariant of Q.[8]

By applying this idea to the notion of equivalence between sets through one-to-one correspondences (the equinumericity relation being an equivalence relation), Kronecker arrives at the idea that numbers, rather than objects associated to totalities of equipotent sets, as in Frege, are invariants associated to equinumericity classes.

Since numbers do not pre-exist their use as invariants, it is a convention, partly arbitrary, partly opportunistic, that will determine their choice. It is a general fact that, if one knows how to choose a particular object x in each equivalence class S, this choice determines a system of invariants for the relation considered.[9] It is therefore "sufficient" to proceed in this way to construct the numbers, and in this fashion one can interpret the construction (posterior to Kronecker) of the numbers starting from the empty set $\emptyset$, which interprets the number 1 as the set $\{\emptyset\}$, the number 2 as the set $\{\emptyset, \{\emptyset\}\}$, and so on.

In children and primitive tribes where counting stops at small numbers, a body counting rule determines the invariant associated with a particular class. Three fingers "are" thus the number three in the sense that, associated with any group of three objects by the process of body counting, they represent for them the number three just as faithfully as for us the conventional sign "3". For us and for mathematically evolved civilizations, it is advisable to choose a system of invariants that is as neutral and devoid of external meanings as possible: the sequence of ordinals lends itself well to this. Kronecker thus justifies his preference for a foundation on ordinals, in the light of the mathematics of his time and the understanding of equinumericity that goes with it.

It is necessary to recognize in this point of view a certain legitimacy. When we count, we certainly do not have in mind all the Fregean sets arranged in classes of equipotence. On the other hand, we know that the process of empirical numbering, which amounts to sequentially choosing elements of the group of objects considered and associating the corresponding ordinals to them, will give us the right result. In practice, we replace the original intuition of dividing all finite sets into equipotence

---

[6]In the air of the time at the end of the nineteenth century, but of which he gives a clear, precise and explicit account.

[7]$x = ux' + vy'$, $y = tx' + wy'$ with $uw - vt = 1$ or -1.

[8]Ideally, in this type of situation one even looks for functions or systems of functions that characterize Q up to equivalence, i.e. such that $f(Q) = f(Q')$ if and only if Q is equivalent to Q'.

[9]Simply define f by $f(y) := x$ for y belonging to S.

classes with a new intuition based on our empirical knowledge of the counting algorithm and on the fact that it associates invariants to the sets (two sets to which the same number is associated by the counting procedure have the same cardinal).

The approach of numbers by the idea of invariants has yet another advantage. It allows us to understand, on the simple and intuitive example of numbers, the essence of the general mathematical process at work in the calculations of Gauss and Lagrange. This process is one of the foundations of one of the leading mathematical disciplines of the twentieth century (the greatest mathematical success of this century according to some authors): algebraic topology. It is worth saying a few words about it, because the original intuition behind it is not so far from questions raised about the emergence of numbers from sensitive intuitions.

One of the first possible ways to approach number naively rests on the principle of individuation of objects, which itself rests on the principle of continuity: an object is individuated when it presents itself as an inseparable whole. In topology, this notion of inseparability is called connectedness, and it is the very first work of topology to count the number of "connected components". A second level consists in calculating the number of holes in a surface (it can be shown that the notion is well defined).[10] These ideas have higher dimensional analogues and give rise to the Betti numbers, Poincaré groups, homotopy groups and homology groups of geometers (Dieudonné [44]).

The key point here, for us, is twofold, beyond the technical details. On the one hand, these calculations and theories, typical of the twentieth century, renew the content of very old ideas about the natural association between number concepts and the principle of geometric individuation of objects. On the other hand, it so happens that the mathematical definition of all these notions ("numbers of connectedness", Betti numbers...) rests exactly on the principle developed by Gauss and Lagrange: two geometrical objects are said to be equivalent if they can be deformed into each other by a continuous transformation, and the numbers calculated by algebraic topology are nothing other than invariants, in the sense of Kronecker, for this notion of equivalence. Poincaré's famous conjecture, recently solved, rests entirely on this logic, but the proof of Fermat's theorem would not exist either without these topological techniques. Let us insist on it: this is a very general principle, on which innumerable advances, among the most important of the last century, are based.

The Kroneckerian approach to numbers thus brings to light new features of mathematical thought, new heuristics and a methodology whose field of action, to be illustrated remarkably on the example of numbers, has a considerable extension. The notion of invariance, in particular, which we have only been able to address here very superficially, is linked to some of the deepest ideas mathematics and the natural sciences have in common. It would be desirable for the philosophy of science to take a greater interest in the future in this type of fundamental idea—a difficult task

---

[10]For details on these notions and metamorphoses of the corresponding mathematical idea of genus, see Popescu-Pampu [98].

where the conceptual history of mathematics joins philosophy and mathematical technique proper.

15.2   Universal Problems

Approaching numbers through constituted mathematical thinking is not limited to the sole consideration of numbers as invariants. Algebra also has its say and, here again, the phenomena that emerge illustrate much more general ideas and strategies, which the twentieth century has largely used to construct some of its most remarkable mathematical theories.

We have already had the opportunity to briefly evoke the algebraic approach to numbering, with Leibniz noting that one can demonstrate that $2 + 2 = 4$ from the recursive definition of numbers: $2 = 1 + 1, 3 = 2 + 1, 4 = 3 + 1$, since then: $2 + 2 = 2 + 1 + 1 = 3 + 1 = 4$. The explicit call to two important properties of integer addition is missing from these calculations to be conclusive, associativity and commutativity,[11] an observation which seems to be due to Johann Schultz in 1789 (Potter [99, p. 59]) and was further developed by Grassmann (1809–1877).

The axiomatic approach to algebra can be developed from there. Today we call a commutative monoid a set S with a zero[12] and an associative and commutative law. Innumerable algebraic systems satisfy the axioms of monoids, starting with all the usual extensions of the set of natural numbers (integers, rational numbers, real numbers, complex numbers...). The set of natural numbers (including 0) plays a special role in this theory, that of a universal object. It can be shown quite easily that if S is an arbitrary commutative monoid and s is any element of S, there is a single morphism f of monoids[13] from the set of natural numbers to S such that $f(1) = s$. This property even characterizes the set of natural integers (including 0).[14] This phenomenon is, again, very general and admits different levels of formulation. In algebra, it corresponds to the notion of free object.[15] In category theory, an even more general presentation can be given,[16] which we will limit ourselves to

---

[11] $x + (y + z) = (x + y) + z, x + y = y + x$.

[12] The neutral element for addition: $0 + x = x + 0 = x$.

[13] I.e. a function preserving the structures: $f(0) = 0$, where we still denote by 0 the neutral element of S, and $f(x + y) = f(x) + f(y)$.

[14] In the sense that any monoid satisfying this property is canonically identified with the set of natural numbers (it is isomorphic to it).

[15] The monoid of natural numbers (including 0) is the free monoid on one generator in the same way that the algebra of real polynomials is the free commutative unitary algebra on one generator over the real field.

[16] The generation of natural numbers from 1 is then given by what is called a left adjoint, a simple reformulation in the language of categories of the property of the set of integers to be a universal object or the free monoid on a generator (all these notions being equivalent). However, the notion of adjoint has a scope that goes far beyond this type of algebraic construction. In its full generality,

illustrating with an elementary example taken from the universe of numbers and which, in spite of its simplicity, exhibits the underlying mathematical mechanisms.

The notion of upper bound is extremely useful and, from the point of view of praxis, was acquired by mankind long before any form of mathematization. Let us consider a simple example: let us fix two numbers, 3 and 5. Then, for any number n greater than or equal to both 3 and 5, there is a number k which is both greater than or equal to 3 and 5 and less than or equal to n. This number is usually not unique (if $n = 8, k = 5, 6, 7, 8$ are all solutions to the problem). However, this problem has a universal formulation: the number n can vary freely in the set of numbers greater than or equal to 3 and 5. The only number k which is the solution of the universal problem (i.e. the only number which is the solution of the problem for all these values of n) is the upper bound of the set 3, 5 (5, in this case).

In the language of categories, an inequality such as $3 \leq 8$ is interpreted as a morphism of 3–8. The property of universality of the upper bound s of two numbers a and b translates into the statement: "For any morphism from a to n and any morphism from b to n, there is a morphism from s to n factoring the previous morphisms" (in the sense that: $a \leq s \leq n$ and $b \leq s \leq n$). It is then said that "s is the coproduct of a and b". The same construction scheme makes it possible to define notions as apparently disparate as logical connectors (and/or), Cartesian products and sums of sets: in the construction that has just been carried out, it is sufficient to replace numbers by objects and inequalities by morphisms of the category considered.[17]

The details of the construction would take us too far away from the problem of the nature of numbers, but it was appropriate to mention it. All this shows how multiform and ramified the conceptual horizon of numbers is. In a sense, these ideas are already in germ in Dedekind, who had glimpsed and demonstrated this universal character of numbers in the context of set theory. His statement of the definition by induction[18] gives indeed, in set theory (and not in an algebraic context), a universal characterization of integers quite similar to the one that can be obtained in the algebraic way. Given an arbitrary function g of a set S to itself and given an element s of S, Dedekind indeed observes that there is a single function f of the sequence of natural numbers to S such that $f(0) = s$ and $f(n+1) = g(f(n))$. Dedekind's result can be interpreted in algebraic terms[19] and can thus be understood as a corollary of the universal property of the set of integers from the point of view of monoid theory.

---

it is, once again, one of the founding principles of contemporary mathematics. See for example Mac Lane [86].

[17] See Patras [90].

[18] Dedekind [36, sect. 126].

[19] By noticing that the maps of S to itself (called the *endomorphisms* of S) form a monoid with respect to composition. One can indeed always compose two maps of S to itself. This composition defines an associative product with neutral element the identity function. The set of iterations $Id = g^0, g, g^2, g^3 \ldots$ of g forms a commutative monoid. There is thus a single function F from the natural numbers to the endomorphisms of S such that $F(0) = Id$, $F(1) = g$, so that $F(n) = g^n$. Setting $f(n) := F(n)(s)$, we find the result of Dedekind.

The universal problem approach to the construction of mathematical objects is of considerable technical, but also speculative, interest. It shows that, in a given mathematical context, certain objects "emerge" spontaneously from the underlying structures. These objects, because of their properties of universality, often capture the essence of the general theory. In the case of numbers, their algebraic universality partly explains their protean character: the game we have played in translating Dedekind's algebraic construction into an algebraic language shows how two a priori quite different approaches to numbers and their fundamental properties can be identified at the cost of a few mathematical manipulations, which, it must be admitted, would be difficult for philosophical discourse to account for.

15.3   Phenomenology of Algebra

Should we, however, give up a philosophical view on the type of mathematical activity at work in these universal phenomena? It is tempting to go beyond them and put on signitive, symbolic, algebraic or categorical activity a reflexive view similar to that which has been taken of the elementary mechanisms of number formation by seeking to identify the cognitive phenomena underlying the corresponding mathematical techniques.

The simplest example of a situation of a mechanical activity is that of counting, which is possible without the need to reactivate the arithmetic meaning of number statements. For example, imagine counting ballots on election night:

> Surely then we can no longer hold together in a single act the successive apprehensions of the members of the multiplicity. Each time there is only a small number that remains in a state of clear distinction. While continually new members are apprehended, others among those who have been apprehended before escape; the acts which present them to consciousness fade ever more into the background of consciousness and vanish altogether. Yet we possess a certain concept of the unity of the whole process. Even if there is only its last, very limited, piece that is actually present before us, we are nevertheless aware of the fact that this piece is not the whole process [...]. With all this, we can construct the symbolic representation of a complete process, which, in any succession, leads to the apprehension of all possible members of the intuitive whole.
>
> Husserl [71]

Husserl abandons here the static conception of number as a collection of units and adopts a Pythagorean and dynamic conception, in the logic of Dedekind or Kronecker's approaches. It is now the process that is intuitively understandable and constitutes the foundation of the activity of numbering, and no longer the totality, understood as such, of the contents of a collection. But how does the passage from isolated moments of a process to the representation of a totality of possible members of an intuitive whole take place? Isn't it that the moments of consciousness involved in the different instants of the numbering process have an invariable structure, which could be detected by a well-conducted phenomenological analysis that would meet Kronecker's understanding of numbers as invariants? And wouldn't this fundamental link between intentional analysis and mathematical methods explain why the mathematician is able to operate intuitively with symbolic and algebraic processes as well as with the intuitive materials provided by the intuition of space or time? Or is it not because the intentionality at work in the different moments of the enumeration process obeys universal rules and structures, and can these structures not be analyzed through a study of the noetic moments of the enumeration activity?

This is the thesis that we will defend: the phenomena addressed by Husserl, in the key pages of the *Philosophy of Arithmetic* that we have just cited, should not be analyzed as relying on a symbolic activity, but on an original way of thinking, whose best mathematical description undoubtedly passes to this day through contemporary algebra and the theory of categories.

Let us therefore take up again step by step the terms of the Husserlian analysis, allowing us a technical analysis of its different moments. Let us assume that we are confronted with a totality that is too large to be apprehended by direct intuition, or even a rather ill-defined totality (such as "the stars visible in the sky"). How is a numbering scheme then formed that will allow us to speak of numbers in the absence of an adequate representation in a naive sense (the representation under a well-defined totality of individuals stripped of their own features and identified with as many units)?

The answer given by Husserl is edifying: the concept of number appears, in such situations, by the sole consideration of the iterative scheme of passing from one element of the totality to another. This scheme is itself essentially ideal: what is important is the possibility of the existence of such a scheme rather than its actual implementation. In the words of Husserl: "And so, with all this, we can construct the symbolic representation of a complete process, which, in any succession (this succession is even indifferent to us), leads to apprehending all the possible members of the intuitive whole."

The possibility of numbering depends therefore on the existence of a universal scheme based on the notion of succession in a given totality. This is, of course, essentially Dedekind's construction, but the same scheme of analysis could be applied to all phenomena involving free objects or categorical adjunctions, provided that one takes the trouble to replace the scheme associated with succession by the scheme typical of the problem under consideration. For example, the algebra of polynomials with integer coefficients is thus nothing other than the mathematical object obtained by considering an indeterminate x and all the expressions obtained from addition and multiplication operations subject to the usual rules (associativity, commutativity, distributivity...). ...: x, $2x$, x^2, $x + 2x^3$ .... As in the case of numbers, this is an ideal process; no one will ever be able to construct all polynomials in this way, but knowledge of the construction process is enough for us to have a very precise intuition of the object thus constructed. Mathematical intuition, at a certain level, is no longer an intuition of objects but of their formation processes and the uses that can be made of them. The canonical example is given by the use of numbers in inductive reasoning—the object of Dedekind's theorem which, in contemporary language, characterizes the set of natural numbers as a solution to a universal problem.

Numbers are thus "natural", like countless mathematical objects, not in that their existence would testify to the problematic presence of ideal numbers in the sky of Ideas of the various neo-Platonisms, but in that their construction, existence and uniqueness derive necessarily from well-posed problems and fundamental cognitive phenomena, some of which admit remarkable mathematical presentations.

Epilogue

Any scientific or philosophical question, when addressed systematically and its implications developed, leads far beyond what might be expected at first sight. The question of the nature of numbers, in its apparent simplicity: what could be simpler, more universal, more elementary! is no exception to the rule and leads to many key ideas in the intellectual history of humanity. The conclusions that emerge from its study are manifold; we will not return here to those concerning the numbers themselves, but rather to general principles that result from them.

A first observation relates to the variability and extent of the questions raised by the very existence of mathematical objects and to their emblematic character, already manifest in Greek philosophy. For us moderns, this variability takes other forms, ranging from logical and foundational issues in mathematics to the problem of meaning; from Frege and Russell to Husserl; from Hilbert to Gödel.

This evolution in the way numbers and mathematics are viewed over the centuries should not conceal another, equally fundamental phenomenon: the persistence throughout the ages of questions and points of view which, although renewed, still retain the mark of their origin. The third man argument and the problem of actual infinity, which are already in Plato, are found in Bolzano and Dedekind in an essentially identical form. The dynamic conception of number, but also the philosophical problems that go with it—What is a concept? What role does thought play in the creation of idealities?—retain all their topicality and regain freshness and even depth to be taken back to their Greek origins and their history.

On the other hand, it is undeniable that the development of science is profoundly changing the way we can and must look at the philosophy of mathematics. The algebraic viewpoint, for example, in all its modernity, brings us in contact with scientific thought mechanisms that were inconceivable before its emergence in the nineteenth century and its deepening throughout the twentieth century. It teaches us that the mathematical object is often constituted by crystallization from well-posed problems, according to a law immanent to the problem itself. This sheds an entirely new light on the problems of mathematical creation by freeing a space of a logic of

F. Patras, *The Essence of Numbers*, Lecture Notes in Mathematics 2278, https://doi.org/10.1007/978-3-030-56700-2

concept formation. The latter is very present in contemporary algebra and category theory but its philosophical implications remain largely unthought of.

The last major lesson that should be emphasized is the relevance of the problem of ontology: how are the objects of our scientific knowledge constituted? Once the formalistic temptation of the 1920s to evacuate ontology and metaphysics from the discourse on science has been overcome, the whole evolution of mathematical philosophy and logic during the twentieth century militates in favour of a revival of these questions on renewed foundations. Phenomenology, post-Gödelian logic, but also, and perhaps above all, mathematics itself, when it manages to take a reflexive look back at the content of its thinking, are all tools for moving in this direction.

References

1. Alexandre d'Aphrodise, in *Alexandri quod fertur in Aristotelis Sophisticos Elenchos commentarium*, ed. by M. Wallies (Berlin, 1898)
2. Aristotle, *On Sophistical Refutations, On Coming-to-be and Passing Away*, transl. E.S. Forster; *On the Cosmos*, transl. D.J. Furley (Heinemann, London, 1955)
3. Aristotle, *The Categories, On Interpretation, Prior Analytics*, transl. H.P. Cook (Harvard University Press, Cambridge, 1962)
4. Aristotle, *Metaphysics. Books M and N*, transl. J. Annas (Clarendon Press, Oxford, 1976)
5. Aristotle, *Catégories*, transl. (French) J. Tricot (Vrin, Paris, 1984)
6. Aristotle, *La Métaphysique*, transl. (French) J. Tricot (Vrin, Paris, 1986)
7. Aristotle, *Metaphysics. Books Γ, Δ and E*, 2nd edn. transl. Ch. Kirwan (Clarendon Press, Oxford, 1993)
8. A. Arnauld, B. Pascal, F. de Nonancourt, *Géométries de Port-Royal*, ed. by D. Descotes (Honoré Champion, Paris, 2009)
9. F. Bailly, G. Longo, *Mathématiques et sciences de la nature. La singularité physique du vivant* (Hermann, Paris, 2006)
10. F. Bergeron, G. Labelle, P. Leroux, *Combinatorial Species and Tree-Like Structures* (Cambridge University Press, Cambridge, 1998)
11. L. Bienvenu, M. Hayrup, Une brève introduction à la théorie effective de l'aléatoire. Gaz. Math. **123**, 35–47 (2010)
12. O. Biermann, *Theorie der analystischen Funktionen* (Teubner, Leipzig, 1887)
13. L. Boi, *Le Problème mathématique de l'espace* (Springer, Berlin, 1995)
14. B. Bolzano, *Paradoxien des Unendlichen* (Leipzig, 1851)
15. J. Boniface, *Hilbert et la notion d'existence en mathématiques* (Vrin, Paris, 2004)
16. N. Bourbaki, *Eléments de mathématiques* (Hermann, then Masson, Paris, 1939–1984)
17. N. Bourbaki, *Théorie des ensembles* (Hermann, Paris, 1970)
18. N. Bourbaki, L'architecture des Mathématiques, in *Les grands Courants de la Pensée mathématique*, ed. by F. Le Lionnais (Hermann, Paris, 1998)
19. R. Brisart (ed.), *Husserl et Frege. Les ambiguïtés de l'antipsychologisme* (Vrin, Paris, 2002)
20. G. Cantor, Über die Ausdehnung eines Satzes aus der Theorie der trigonometrischen Reihen. Math. Ann. **5**, 123–132 (1872)
21. G. Cantor, Über eine Eigenschaft des Inbegriffes aller reellen algebraischen Zahlen. J. Reine Angew. Math. **77**, 258–263 (1874)
22. G. Cantor, Ein Beitrag zur Mannigfaltigkeitslehre. J. Reine Angew. Math. **84**, 242–258 (1878)

© The Editor(s) (if applicable) and The Author(s), under exclusive license
to Springer Nature Switzerland AG 2020
F. Patras, *The Essence of Numbers*, Lecture Notes in Mathematics 2278,
https://doi.org/10.1007/978-3-030-56700-2

23. G. Cantor, Über unendliche lineare Punktmannigfaltigkeiten, 3. Math. Ann. **20**, 113–121 (1882)
24. G. Cantor, *Grundlagen einer Mannigfaltigkeitslehre. Ein mathematisch-philosophischer Versuch in der Lehre des Unendlichen* (Teubner, Leipzig, 1883)
25. G. Cantor, *Gesammelte Abhandlungen zur Lehre vom Transfiniten* (C.E.M. Pfeffer, Halle, 1890)
26. G. Cantor, Beiträge zur Begründung der transfiniten Mengenlehre. Math. Ann. **46–49**, 481–512; 207–246 (1895–1897)
27. P. Cartier, *Vie et mort de Bourbaki. Notes sur l'histoire et la philosophie des mathématiques I* (IHES, Paris, 1997)
28. E. Cassirer, *Individuum und Kosmos in der Philosophie der Renaissance* (B.G Teubner, Leipzig, 1927)
29. P. Cassou-Noguès, *Hilbert* (Belles Lettres, Paris, 2001)
30. É.B. de Condillac, La Langue des calculs, in *Œuvres philosophiques* (PUF, Paris, 1948)
31. A. Connes, A. Lichnerowicz, M.-P. Schützenberger, *Triangle de pensées* (Odile Jacob, Paris, 2000)
32. L. Corry, *Modern Algebra and the Rise of Mathematical Structures* (Birkhäuser, Basel, 1996)
33. L. Couturat, Sur une définition logique du nombre. Rev. Metaphys. Morale **8**, 23–36 (1900)
34. A. Dahan Dalmedico, L'étoile imaginaire a-t-elle immuablement brillé? Le nombre complexe et ses différentes interprétations dans l'œuvre de Cauchy, in *Le Nombre, une hydre à n visages*, ed. by D. Flament (FMSH, Paris, 1987)
35. Ch. de Bovelles, *Le Sage*, (1509). Reprinted in Cassirer, *Individu et Cosmos dans la philosophie de la Renaissance* (Minuit, Paris, 1983)
36. R. Dedekind, *Was sind und was sollen die Zahlen?* (Vieweg, Brunswick, 1888)
37. S. Dehaene, *The Number Sense* (Oxford University Press, Oxford, 1997)
38. P. Dehornoy, Cantor et les infinis. Gaz. Math. **121**, 28–46 (2009)
39. R. Descartes, *La Géométrie* (Jean Maire, Leyde, 1637)
40. R. Descartes, *Discours de la méthode* (Jean Maire, Leyde, 1637)
41. R. Descartes, *Discourse on Method, and Metaphysical Meditations*, translated by G.B. Rawlings (Walter Scott Publishing, London, 1901) (republished by Franklin Classics, 2018)
42. J. Dieudonné, *Cours de géométrie algébrique*, vol. 2 (Puf, Paris, 1974)
43. J. Dieudonné, *Abrégé d'histoire des mathématiques* (Hermann, Paris, 1978)
44. J. Dieudonné, *A History of Algebraic and Differential Topology, 1900–1960* (Birkhäuser, Bâle, 1989)
45. H.-D. Ebbinghaus et al., *Numbers*. Graduate Texts in Mathematics, vol. 123 (Springer, Berlin, 1990)
46. Euclid, *The thirteen Books of Euclid's Elements*, transl. T.L. Heath, Cambridge (1908)
47. S. Feferman, The impact of the incompleteness theorem on mathematics. Not. AMS **53**(4), 434–439 (2006)
48. B. de Fontenelle, in *Éléements de la géométrie de l'infini. Suite des Mémoires de l'Académie royale des sciences* (Paris, 1727)
49. G. Frege, *Begriffschrift: eine der arithmetischen nachgebildete Formelsprache des reinen Denkens* (Louis Nebert, Halle, 1879)
50. G. Frege, *Grundlagen der Arithmetik* (Wilhelm Koebner, Breslau, 1884)
51. G. Frege, Über Begriff und Gegenstand. Vjschr. Wiss. Philos. **16**, 192–205 (1892)
52. G. Frege, *Grundgesetze der Arithmetik, begriffsschriftlich abgeleitet*, vol. 1 (Iena, 1893)
53. G. Frege, Diary entries on the concept of number, (1924–25), in *Posthumous Writings* (Basil Blackwell, Oxford, 1979)
54. G. Frege, *Posthumous Writings* (Basil Blackwell, Oxford, 1979)
55. H. Gericke, *Geschichte des Zahlbegriffs* (Bibliographisches Institut AG, Mannheim, 1970)
56. G. Godefroy, *L'Aventure des nombres* (Odile Jacob, Paris, 1997)
57. H. Hankel, *Theorie der complexen Zahlsysteme* (Leipzig, 1867)
58. M. Heidegger, *Kant und das Problem der Metaphysik* (Frankfurt, 1965)

59. Heijenoort J. van (éd.), *From Frege to Gödel, A Source Book in Mathematical Logic, 1879–1931* (Harvard University Press, Cambridge, 1967)
60. D. Hilbert, *Grundlagen der Geometrie* (B.G. Teubner, Leipzig, 1899)
61. D. Hilbert, On the foundations of logic and arithmetic. Monist **15**(3), 338–352 (1905)
62. D. Hilbert, Die Grundlegung der elementaren Zahlenlehre. Math. Ann. **102**, 485–486 (1930)
63. D. Hume, *An Enquiry Concerning Human Understanding* (1748)
64. E. Husserl, Le concept de l'arithmétique générale, (1890), Articles sur la Logique, transl. (French) J. English (Puf, Paris, 1975)
65. E. Husserl, Sur la logique des signes (sémiotique), (1890), Articles sur la Logique, transl. (French) J. English (Puf, Paris, 1975)
66. E. Husserl, Psychologische Studien zur elementaren Logik. Philosophische Monatshefte **30**, 159–191 (1894)
67. E. Husserl, *Cartesianische Meditationen und Pariser Vorträge* (Nijhoff, Den Haag, 1950)
68. E. Husserl, *Ideen zu einer reinen Phänomenologie und phänomenologischen Philosophie,* Husserliana (Hua) III, Hua IV, Hua V (Nijhoff, Den Haag, 1952)
69. E. Husserl, *Der Ursprung der Geometrie,* Hua VI (Nijhoff, Den Haag, 1954)
70. E. Husserl, *Die Krisis der europäischen Wissenschaften und die transzendentale Phänomenologie,* Hua VI (Nijhoff, Den Haag, 1954)
71. E. Husserl, *Philosophie der Arithmetik: Logische und psychologische Untersuchungen,* 2nd edn. Hua XII (Nijhoff, Den Haag, 1970)
72. E. Husserl, *Formale und Transzendentale Logik: Versuch einer Kritik der logischen Vernunft* (Nijhoff, Den Haag, 1974)
73. E. Husserl, *Logische Untersuchungen,* Hua XVIII, Hua XIX/1, Hua XIX/2 (Nijhoff, Den Haag, 1975–84)
74. E. Husserl, *Studien zur Arithmetik und Geometrie,* Hua XXI (Nijhoff, Den Haag, 1983)
75. G. Ifrah, *Histoire universelle des chiffres* (R. Laffont, Paris, 1994)
76. M. Kaltenmark, *Lao Tseu et le taoïsme* (Seuil, Paris, 1965)
77. E. Kant, *Versuch, den Begriff der negativen Größen in die Weltweisheit einzuführen* (1763)
78. E. Kant, *Untersuchung über die Deutlichkeit der Grundsätze der natürlichen Theologie und der Moral* (1764)
79. E. Kant, *Kritik der reinen Vernunft* (1781–87)
80. L. Kronecker, Sur le concept de nombre en mathématiques, ed. and French transl. J. Boniface et al.. Rev. Histoire Math. **7**, 207–275 (2001)
81. Lecourt, D. (ed.), *Dictionnaire d'histoire et de philosophie des sciences* (Puf, Paris, 1999).
82. J. Locke, *An Essay Concerning Human Understanding* (1690)
83. G. Longo, On the proofs of some formally unprovable propositions and prototype proof theory in type theory, Invited Lecture, *Types for Proofs and Programs* (Durham, 2000)
84. G. Longo, Prototype proofs in type theory. Math. Logic Q. **46**(3), 257–266 (formely: Zeitscrift f. Math. Logik u. Grundlagen der Math.) (2000)
85. G. Longo, Interfaces de l'incomplétude, in *La Matematica,* vol. 4 (Einaudi, 2010)
86. S. Mac Lane, *Categories for the Working Mathematician* (Springer, Berlin, 1971)
87. E. Nagel, J.R. Newman, *Gödel's Proof.* Revised Edition (New York University Press, New York, 2001)
88. M. Ohm, *Kritische Beleuchtung der Mathematik überhaupt und der Euklidischen Geometrie insbesondere* (Berlin, 1819)
89. F. Patras, *La Pensée mathématique contemporaine* (Puf, Paris, 2001)
90. F. Patras, Phénoménologie et théorie des catégories, in *Geometries of Nature, Living Systems and Human Cognition,* ed. by L. Boi (World Scientific, Singapore, 2005)
91. F. Patras, Pourquoi les nombres sont-ils "naturels"? in *Rediscovering Phenomenology,* ed. by L. Boi, P. Kerszberg, F. Patras (Springer, Berlin, 2007)
92. F. Patras, Objets et idéalités dans les mathématiques contemporaines. Études Platoniciennes IX [Platon aujourd'hui] **2012**, 47–61 (2012)
93. G. Peano, *Formulaire de mathématiques,* vol. 4 (Bocca et Clausen, Turin, 1895–1903)
94. Plato, *Parménide,* transl. (French) E. Chambry (Garnier, Paris, 1967)

95. Plato, *Parmenides*, transl. R.E. Allen (Yale University Press, New Haven, 1997)
96. Plotinus, *Traité sur les nombres, Ennéade VI 6*, transl. (French) J. Bertier et al. (Vrin, Paris, 1980)
97. Plotinus, Ennead VI 6 (On Numbers), in *Plotinus, vol. VII*, Greek text with English translation by A.H. Armstrong (Loeb Classical Library, Cambridge, 1988)
98. P. Popescu-Pampu, *What is the Genus?* Lecture Notes in Mathematics. History of Mathematics Subseries, vol. 2162 (Springer, Berlin, 2016)
99. M. Potter, *Reason's nearest Kin* (Oxford University Press, Oxford, 2002)
100. B. Russell, Mathematical logic as based on the theory of types. Am. J. Math. **30**(3), 222–262 (1908)
101. K.H. Scholte, Des nombres complexes aux systèmes hypercomplexes, in *Le Nombre, une hydre à n visages*, ed. by D. Flament (FMSH, Paris, 1987)
102. G. Schubring, Ruptures dans le statut mathématique des nombres négatifs. petit x, n° **12**, 5–32 (1986)
103. G. Schubring, L'interaction entre les débats sur le statut des nombres négatifs et imaginaires et l'émergence de la notion de segment orienté, in *Le Nombre, une hydre à n visages*, ed. by D. Flament (FMSH, Paris, 1987)
104. G. Schubring, *Conflicts between Generalization, Rigor and Intuition* (Springer, Berlin, 2005)
105. J. Sesiano, The appearance of negative solutions in mediaeval mathematics. Arch. Hist. Exact Sci. **32**(4), 105–150 (1985)
106. A. Sokal, J. Bricmont, *Impostures intellectuelles* (Odile Jacob, Paris, 1997)
107. R. Thom, *Paraboles et Catastrophes* (Flammarion, Paris,1983)
108. A. Weil, *Souvenirs d'apprentissage* (Springer, Berlin, 1991)
109. H. Weyl, *Das Kontinuum*, (1918), transl. S. Pollard and Th. Bole: *The Continuum, A Critical Examination of The Foundations of Analysis*, reimpr. (Dover, 1994)
110. H. Weyl, *Symmetry* (Princeton University Press, Princeton, 1952)
111. A.N. Whitehead, B. Russell, *Principia Mathematica*, vol. 3 (Cambridge University Press, Cambridge, 1910–1913)
112. L. Wittgenstein, *Tractatus logico-philosophicus* (Basil Blackwell, Oxford, 1959)
113. L. Wittgenstein, *Lectures on the Foundations of Mathematics, Cambridge, 1939*, ed. by C. Diamond (University of Chicago Press, Chicago, 1939)

Index

LECTURE NOTES IN MATHEMATICS  Springer

Editors in Chief: J.-M. Morel, B. Teissier;

Editorial Policy

1. Lecture Notes aim to report new developments in all areas of mathematics and their applications – quickly, informally and at a high level. Mathematical texts analysing new developments in modelling and numerical simulation are welcome.

   Manuscripts should be reasonably self-contained and rounded off. Thus they may, and often will, present not only results of the author but also related work by other people. They may be based on specialised lecture courses. Furthermore, the manuscripts should provide sufficient motivation, examples and applications. This clearly distinguishes Lecture Notes from journal articles or technical reports which normally are very concise. Articles intended for a journal but too long to be accepted by most journals, usually do not have this "lecture notes" character. For similar reasons it is unusual for doctoral theses to be accepted for the Lecture Notes series, though habilitation theses may be appropriate.

2. Besides monographs, multi-author manuscripts resulting from SUMMER SCHOOLS or similar INTENSIVE COURSES are welcome, provided their objective was held to present an active mathematical topic to an audience at the beginning or intermediate graduate level (a list of participants should be provided).

   The resulting manuscript should not be just a collection of course notes, but should require advance planning and coordination among the main lecturers. The subject matter should dictate the structure of the book. This structure should be motivated and explained in a scientific introduction, and the notation, references, index and formulation of results should be, if possible, unified by the editors. Each contribution should have an abstract and an introduction referring to the other contributions. In other words, more preparatory work must go into a multi-authored volume than simply assembling a disparate collection of papers, communicated at the event.

3. Manuscripts should be submitted either online at www.editorialmanager.com/lnm to Springer's mathematics editorial in Heidelberg, or electronically to one of the series editors. Authors should be aware that incomplete or insufficiently close-to-final manuscripts almost always result in longer refereeing times and nevertheless unclear referees' recommendations, making further refereeing of a final draft necessary. The strict minimum amount of material that will be considered should include a detailed outline describing the planned contents of each chapter, a bibliography and several sample chapters. Parallel submission of a manuscript to another publisher while under consideration for LNM is not acceptable and can lead to rejection.

4. In general, **monographs** will be sent out to at least 2 external referees for evaluation.

   A final decision to publish can be made only on the basis of the complete manuscript, however a refereeing process leading to a preliminary decision can be based on a pre-final or incomplete manuscript.

   Volume Editors of **multi-author works** are expected to arrange for the refereeing, to the usual scientific standards, of the individual contributions. If the resulting reports can be forwarded to the LNM Editorial Board, this is very helpful. If no reports are forwarded or if other questions remain unclear in respect of homogeneity etc, the series editors may wish to consult external referees for an overall evaluation of the volume.

5. Manuscripts should in general be submitted in English. Final manuscripts should contain at least 100 pages of mathematical text and should always include

   – a table of contents;
   – an informative introduction, with adequate motivation and perhaps some historical remarks: it should be accessible to a reader not intimately familiar with the topic treated;
   – a subject index: as a rule this is genuinely helpful for the reader.
   – For evaluation purposes, manuscripts should be submitted as pdf files.

6. Careful preparation of the manuscripts will help keep production time short besides ensuring satisfactory appearance of the finished book in print and online. After acceptance of the manuscript authors will be asked to prepare the final LaTeX source files (see LaTeX templates online: https://www.springer.com/gb/authors-editors/book-authors-editors/manuscriptpreparation/5636) plus the corresponding pdf- or zipped ps-file. The LaTeX source files are essential for producing the full-text online version of the book, see http://link.springer.com/bookseries/304 for the existing online volumes of LNM). The technical production of a Lecture Notes volume takes approximately 12 weeks. Additional instructions, if necessary, are available on request from lnm@springer.com.

7. Authors receive a total of 30 free copies of their volume and free access to their book on SpringerLink, but no royalties. They are entitled to a discount of 33.3 % on the price of Springer books purchased for their personal use, if ordering directly from Springer.

8. Commitment to publish is made by a *Publishing Agreement*; contributing authors of multiauthor books are requested to sign a *Consent to Publish form.* Springer-Verlag registers the copyright for each volume. Authors are free to reuse material contained in their LNM volumes in later publications: a brief written (or e-mail) request for formal permission is sufficient.

Addresses:
Professor Jean-Michel Morel, CMLA, École Normale Supérieure de Cachan, France
E-mail: moreljeanmichel@gmail.com

Professor Bernard Teissier, Equipe Géométrie et Dynamique,
Institut de Mathématiques de Jussieu – Paris Rive Gauche, Paris, France
E-mail: bernard.teissier@imj-prg.fr

Springer: Ute McCrory, Mathematics, Heidelberg, Germany,
E-mail: lnm@springer.com

Printed in the United States
By Bookmasters